Physics and Chemistry in Space
Volume 3

Edited by
J. G. Roederer, Denver, and J. Zähringer, Heidelberg

Editorial Board:
H. Elsässer, Heidelberg · G. Elwert, Tübingen
L.G. Jacchia, Cambridge, Mass. · J.A. Jacobs, Edmonton
P. Meyer, Chicago, Ill. · N.F. Ness, Greenbelt, Md.
W. Nordberg, Greenbelt, Md. · W. Riedler, Graz
J.W. Warwick, Boulder, Colo.

I. Adler · J. I. Trombka

Geochemical Exploration of the Moon and Planets

With 129 figures

Springer-Verlag New York Heidelberg Berlin 1970

ANNEXE DE LA BIBLIOTHÈQUE
UNIVERSITAS OTTAVIENSIS
uOttawa
LIBRARY ANNEX

I. Adler · J. I. Trombka

National Aeronautics and Space Administration Goddard Space Flight Center
Greenbelt, Maryland, USA

This work is subject to copyright. All rights are reserved, whether the whole or part of the material is concerned specifically those of translation, reprinting, re-use of illustrations, broadcasting, reproducing by photocopying machine or similar means, and storage in data banks.
Under § 54 of the German Copyright Law where copies are made for other than private use, a fee is payable to the publisher, the amount of the fee to be determined by agreement with the publisher.

© by Springer-Verlag Berlin · Heidelberg 1970. Library of Congress Catalog Card Number 78-127039.
Printed in Germany.

The use of general descriptive names, trade names, trade marks etc. in this publication, even if the former are not especially identified, is not to be taken as a sign that such names, as understood by the Trade Marks and Merchandise Marks Act, may accordingly be used freely by anyone.
Title No. 3212.

QC
801
.P46
#3
1970

Preface

This book presents a review of those efforts that have been and are being made to determine the geochemical composition of the moon and planets. The authors have attempted to present both a review as well as their philosophy about the development of flight experiments for geochemical studies. Their basic premise is that such flight experiments should emphasize the scientific objectives and a total systems approach to meeting these objectives, involving the analytical device, data handling and data interpretation. While the above seems reasonably obvious, many proposals of experiments often tend to begin with an instrument with too little concern about the constraints imposed and whether the data that can be obtained are sufficiently useful to meet the scientific objectives.

This book covers the accomplishments in space science exploration, bearing on the history and composition of the solar system. It also covers the rationale behind the lunar and planetary exploration program.

The latter part of the book is concerned with future plans for lunar and planetary exploration instrumentation and techniques in various stages of development. There is an exposition of the methods of remote analysis of the moon and planets, including some concepts developed by the authors as a result of their long term involvement with the space program, from its early inception to the present day preparation for remote geochemical analysis in the Apollo, Mariner and Viking missions.

Although the planet earth has not been mentioned explicitly, the authors believe that the instrumental methods and the analytical procedure describe herein are equally promising in its exploration.

This text would not have been possible without the cooperation of a large number of individuals and institutions. We gratefully acknowledge the use of information supplied by them that does not always appear in the open literature.

We would especially like to acknowledge the following individuals for their help and comments; Dr. J. Arnold and Dr. L. Peterson from the University of California, San Diego; Dr. A. Metzger from J. P. L.; Dr. R. Allenby, Dr. M. Molloy and Mr. R. Bryson of NASA Headquarters; Dr. J. Reed of IITRI; Dr. R. Caldwell and Dr. W. Mills of Socony Mobil; Dr. J. Waggoner of the Lawrence Radiation Laboratory; Mr. R. Schmadebeck of the Goddard Space Flight Center, Dr. F. Senftle of the U.S.G.S.; and finally to Dr. H. Gursky, Dr. P. Gorenstein and Dr. J. Carpenter of American Science and Engineering.

Contents

Chapter 1: Introduction

In the summer of 1965 a distinguished panel of scientists, under the auspices of the Space Science Board of the National Academy of Sciences, met to consider and formulate future plans for lunar and planetary exploration. The results of this study were published under the title "Space Research-Directions for the Future" (1965). To members of the lay community as well as scientists, concerned about the rationale behind such a concerted effort, one cannot do better than to quote the opening statement of the published report. "The exploration of the solar system bears on three central scientific problems of our time: the origin and evolution of our earth, sun and planets; the origin and evolution of life; and the dynamic processes that shape mans terrestrial environment. Not only do these problems bear directly on man's place in the universe but they are closely related to the question of the origin of the universe as well."

The working group of the Space Science Board recognized the obvious need to plan and execute missions of lunar and planetary exploration in a fashion that would yield maximum scientific data. As a very important first step it was essential to review the existing knowledge of the solar system and to identify the significant problems, keeping in mind the limitations imposed by available resources. One very interesting matter to emerge from the deliberations of the working group was a preliminary ranking of the objects in the solar system in order of importance for study, with particular regard to their relevance to an understanding of

Table 1.1. *Ranking of objects of the solar system for study*

Object	Rank
Mars	1
Moon and Venus	2, 3
Major planets	4
Comets and asteroids	5
Mercury	6
Pluto	7
Dust	8

our Earth, the solar system and the origin of life. This ranking is shown in Table 1.1 although a very great emphasis was placed on its preliminary nature. It was fully appreciated that these rankings could easily change with new knowledge as time went on.

It is instructive to look here at the reasons for this ranking. Mars was ranked first because its study is closely tied to each of the three central themes ennunciated above. It is obvious that even at the present stage of our knowledge it represents a body with properties intermediate between the Earth and the Moon.

The interest in the Moon is equally obvious. Here we have the Earth's nearest neighbor. It is an accepted fact that those erosional processes with which we are familiar on the surface of the Earth are missing on the Moon although it is agreed that errosional processes due to meteorite infall and solar flux bombardment are present. Still the material making up the Moon may be the material matter of our solar system and thus an important clue to the early history of our solar system.

The high ranking for Venus comes from a number of reasons. Although preliminary measurements lead to a conclusion of high surface temperatures, there is still great interest in the existence of life forms. For example there is the possibility of elevated topography and lower temperatures and of suspended life forms in the atmosphere. Venus is also a sister planet to the Earth, with a dense atmosphere which is of meteorological importance. Finally some knowledge of the interior of Venus and it's rotational and dynamical characteristics can have considerable bearing on understanding the evolution of the solar system.

The major planets such as Jupiter and Saturn were also accorded a high priority because of the possibility that the chemical composition might be representative of the primitive solar nebula from which the solar system evolved as a part. Not overlooked was the possible formation of prebiotic molecules in the upper atmosphere as an interaction between the reducing upper atmosphere and ultraviolet radiation.

The position of comets and asteroids in the table follows from the likelihood that they contain matter from interstellar space or again the primordial material for the solar system.

Finally the unusual orbits of Mercury and Pluto make them interesting in their relationship to the origin of the solar system.

The low priority given to the study of cosmic dust comes not so much from a low interest quotient but rather from the fact that the methods for study are different and such studies are in progress as parts of other missions.

In addition to establishing a priority list the report of the Space Science Board goes on to list many of the scientific questions to be answered by a program of exploration of the various members of the solar system. These are briefly discussed below.

Mars

Exobiology: Is there or has there ever been any kind of life on Mars and if so what are the chemical processes that support this life? If life is absent and the evidence is that it never did exist on the planet, then are there proto-organic molecules from which life could conceivably evolve?

Differentiation: Does Mars like the Earth have a core, mantle and crust and does the planet's composition vary both vertically and horizontally? If the planet is differentiated are there still ongoing geological processes or did this differentiation take place during the origin of the planet? To answer some of these questions would require refined measurements of the mass, shape and moment of inertia of the planet as well as a study of seismicity and heat flow measurements. Furthermore how active a planet is Mars now and how active has it been from a volcanic, seismic or tectonic point of view?

Composition: The composition of Mars physically, chemically and mineralogically is of great interest because it bears on the question of the origin of Mars as well as the other members of the solar system. A comparison with the Earth is intriguing. To answer these questions would require an analysis of the volcanic rocks and gaseous atmospheres, sediments and isotopes. The role of water and the existence of fossil oceans would be of particular significance.

History: As on the earth the history of the major events on Mars must be recorded in the rocks and the time scale of the events must be understood. To do this would require remote reconnaisance and geological mapping by imaging devices. Thus one could hope to learn about the sequence of superposition of various strata and their relationship to the major events in this history of Mars.

Atmospheric dynamics: What is the nature of the circulation of the Martian atmosphere and what are its driving forces? Is there a relationship between the large scale circulation and the thermal structure as well as the observed cloud forms and the violet layer on Mars? Are there actually dust storms and what is the transport mechanism? What sort of circulation patterns exist on the mesoscale and what occurs at the surface boundary layer? How do these motions at the various scales, if they do exist, affect the transport of water vapor in the atmosphere and across the equator? The answer to these questions, obtained from instrumented landers, combined with existing theories will certainly increase the knowledge of the dynamic forces operating on planetary atmospheres and go a long way towards explaining the surface geology and biology of Mars.

The Moon

The scientific questions about the Moon are divided into three broad categories: the interior, the surface and it's history.

Structure and processes of the lunar interior: Is the Moon radially symmetrical like the earth and like the earth is it macroscopically differentiated into a core, mantle and crust? Does the Moon's geometric shape depart from fluid equilibrium and is there a fundamental difference in the history and morphology of the Moon's averted face compared to the side we see? Since the internal energy regime is of great significance, what is the nature of the heat flow through the lunar surface and what are the sources? Is there active volcanism and does the Moon have its own internally produced magnetic field?

Composition, structure and processes of the lunar surface: What is the average composition of the rocks on the lunar surface, does it vary from place to place and do volcanic rocks appear? What are the forces that shaped the Moon's surface? Are there condensed volatiles and is there any evidence of organic or proto-organic materials on or near the surface?

History of the Moon: What is the age of the Moon? What are the ages of the stratigraphic units and of the oldest exposed material? Is there evidence of exposed primordial material? What is the history of the dynamical interaction between the Earth and the Moon? What is the history of cosmic ray exposure of the lunar surface material and is there an indication of remanent magnetic fields to be found in the lunar rocks?

Venus

Venus continues to be one of the most enigmatic of the planets. The fact that it is completely cloud covered introduces extreme limitations on what can be learned about the surface. There is still uncertainty about the orbit although radar observations indicate a retrograde rotation. Infrared studies indicate the possibility that the clouds may be ice crystals but the complete coverage of the planet is suggestive of the unlikely possibility that there is no downward motion of the atmosphere.

Spectral absorption lines of water vapor and carbon dioxide have been observed at temperatures above 200 °K and at a few atmospheres pressure but because of the highly scattering clouds it is not possible to determine at which level the spectral lines are generated. Radio emission measurements at wavelengths from 3 mm to 40 cm suggests surface temperatures of near 600 °K. A significant question therefore concerns the surface and whether it is solid or liquid? In addition do large topographic features such as high mountains exist on which conceivably some form of life might survive? Perhaps some form of life has developed which is suspended in the very dense atmosphere at high altitudes?

The Outer Planets

By the outer planets one refers to those beyond Mars—Jupiter, Saturn, Uranus, Neptune and Pluto. The first four are called the major planets which are characteristically of large total mass but small mean density. Pluto on the other hand is known to be small with a density comparable to those of the inner planets such as the Earth, Mars, Venus and Mercury.

Jupiter is an enormous planet containing about two-thirds of the mass of the planetary system and most of its angular momentum. It is a very active planet with an atmosphere in continual motion and exhibiting a strong equatorial current. Its magnetic field is very strong and the planet is surrounded by Van Allen type radiation belts. Strong non-thermal radiowaves have been observed. Among the interesting questions is the nature of the Great Red Spot which is unique to Jupiter. Other questions relate to the composition of the planet, atmospheric dynamics, the magnetic field and the radiation belts. With regard to the composition one would like to determine whether the composition of Jupiter is like that of the Sun. Is there a solid interface between the planet and its atmosphere? What is the nature of the Jovian clouds? Is Jupiter more like a planet or a star?

As for the other major planets is the composition of Saturn, Uranus and Neptune non-solar and is Pluto more like the inner planets? One interesting question that arises is whether or nor some from of life or prebiotic material exists on these outer planets?

It is obvious that the questions listed above represent an extraordinary program of exploration and one that would require many generations to bring to fruition. It is also obvious that such a program considered in totality would require a large part of a nations resources. What the Space Science Board has done is to provide scientific objectives around which to build a program for exploring the members of our solar system.

Achievements in Planetology

Having listed many of the questions about our solar system, it is exciting to examine the literally astonishing progress that has already been made since the inception of the space program. Because of the limitations of space and this book's purview we will look only at those advances in planetology defined as the study of the condensed material of the solar system. Although, one must mention in passing that there have been spectacular strides in other areas such as space astronomy, ionospheres, radio physics, particles and fields, solar physics and planetary atmospheres.

Figure 1.1 summarizes the lunar and planetary missions flown since 1958. These include flights to the Moon, Mars and Venus; some as fly-bys, a number as hard landing probes and most recently a number of soft landing probes to the Moon and Venus. Two facts stand out clearly: there has been and increase in frequency of launches with time and an increasing proportion of successes.

A brief review of the progress in lunar and planetary studies is presented below. These are based on the excellent summaries found in NASA SP-155(1967) and SP-136 (1966).

Progress in Lunar Studies Since 1958

Under the impetus of the space program the Moon has been examined in greater detail since 1958 than in all the prior years of recorded history. This has resulted in detailed maps of the equatorial regions. In this period a wave of excitement was set off by the observation of unusual luminescence around the Crater Kepler and visual changes of color near Aristarchus.

I.R. temperature measurements of the cooling rate of the lunar surface after sundown and during eclipse have shown that some areas cool at a slower rate than others. In some instances these were craters and in other mare areas.

In July of 1964 the United States Ranger VII space probe impacted the Moon in the neighbourhood of Mare Cognitum. During the course of its nearly vertical descent it transmitted some 4000 high resolution photographs. The images showed that craters are the most dominant photographic features of the lunar surface. For example in the region Mare Cognitum, craters larger than 1 m in diameter were observed to cover over 50% of the surface and often these larger craters have superimposed smaller craters.

LUNAR AND PLANETARY FLIGHT RECORD

MISSION | 1958 | 1959 | 1960 | 1961 | 1962 | 1963 | 1964 | 1965 | 1966 | 1967 | 1968

PIONEER II
LUNA I
PIONEER IV
LUNA II
LUNA III
PIONEER P3
PIONEER P30
PIONEER P31
SPUTNIK VIII
RANGER III
RANGER IV
MARINER I
MARINER II -E
RANGER V
MARS I
LUNA IV
RANGER VI
RANGER VII
MARINER III
MARINER IV -E
ZOND II
RANGER VIII
RANGER IX
LUNA V
LUNA VI
ZOND III
VENERA II
VENERA III -E
SURVEYOR I
LUNAR ORBITER I
SURVEYOR II
LUNAR ORBITER II
LUNAR ORBITER III
SURVEYOR III
LUNAR ORBITER IV
MARINER V
VENERA IV -E
SURVEYOR IV
EXPLORER XXXV
LUNAR ORBITER V
SURVEYOR V
SURVEYOR VI
SURVEYOR VII

LEGEND:

	UNITED STATES		USSR	
	SUCCESS	FAILURE	SUCCESS	FAILURE
MOON	▲	△	▲	△
MARS	■	□	■	□
VENUS	●	○	●	○
-E PLANETARY ENCOUNTER				

Figure 1.1 Summary of lunar and planetary missions

The Ranger VIII and Ranger IX flights were similarly successful in providing detailed photographs of excellent quality. To summarize the outstanding scientific contributions of the Ranger probe series one can list some of the conclusion drawn by a number of investigators such as Shoemaker, Kuiper et al., Urey etc.:

1. Meteoritic impact as well as volcanism have played major roles in shaping the Moon's topography and lithology.
2. The fine structure of the lunar surface is still undergoing and has undergone shaping by internal processes such as volcanic activity and external processes such as mass wasting, faulting and meteoritic impact.
3. The pre Ranger ideas of the lunar surface topography as gently rolling subdued surfaces with occasional craters and steep slopes overlain with unconsolidated material was essentially correct.

In July of 1965 the USSR launched an automated, interplanetary station Zond 3 which was used, at least in part to photograph the far side of the Moon. One of the most interesting findings was a confirmation of previous observations made in 1959 by Luna 3, that the Moon's far side has no large maria (Lipskiy et al., 1966). In addition the northern and southern "continental shields" visible in the foward hemisphere are continuous around the far side. Zond 3 did demonstrate however that there are large circular depressions on the continental farside, some hundreds of kilometers in size which are mare like. Thus differences between the earthside and farside are not in the absence of mare basins but rather mare material. In the main the flooded maria appear to be concentrated on the Moon's foward of earth side face. The interpretation offered by Lipskiy et al. is that this is strong evidence against the formation of the Moon's maria by the impact of planetesimals which in their view would have resulted in random distribution. They state that the concentration of the maria on the earthward side is either a chance occurrence or perhaps due to the deceleration of the Moon's rotation by the Earth.

A historic event in 1966 was the successful soft landing of the USSR Luna 9 on the lunar surface. The spacecraft carrying a 220 pound instrumental package part of which was a TV camera touched down near the western shore of Oceanus Procellarum. Although the position of the vehicle could not be located with sufficient accuracy to determine whether it had landed on mare or highland terrain, the Russian scientists did publish conclusions about the surface material based both on the photographs and the behavior of the spacecraft. The evidence they felt, ruled against the existence of "fairy castle" structures or fine dust. Vinogradov's (1966) interpretation of the data was that the surface was a basaltic lava, modified however by meteoritic impact, solar radiation, temperature fluctuation etc. There were also a number of published reports by scientists outside of Russia interpreting the Luna 9 photographs. Fielder (1966) et al. suggested that the terrain was vesicular lava. Kuiper et al. (1966) reported that the material appeared to be solid igneous rock such as basalt. Others such as Gault et al. (1966) proposed that the area in the vicinity of the of the lander was probably a non or weakly cohesive fragmental material. This latter point of view has been borne out by subsequent Surveyor and Luna probes about which more will be said below.

In addition to the Luna 9 observations of the surface some measurements were obtained of the background radiation. The figures reported were 30 millirads per day due mainly to cosmic rays and partially to the Moon itself.

Following Luna 9 the Russians succeeded in placing a spacecraft into lunar orbit. This particular probe was identified as Luna 10 and carried a 540 pound payload for performing a variety of physical measurements such as magnetic field strength, cosmic ray, gamma ray and particle intensity flux.

Luna 10 produced exciting results in a number of areas. Cosmic ray measurements, made using gas discharge counters and particle traps gave results in agreement with those from Luna 9. On four separate occasions the scientists observed an increase in counting rate which has been attributed to the spacecraft crossing the boundaries of the Earth's magnetosphere. These Luna 10 results demonstrated that the magnetosphere was to be found at least 300000 km from the Earth. As pointed out in the review of Walter and Lowman (1967) the importance of this discovery is that in any study of the Moon's surface one must look for an enhancement of the solar radiation flux due to the Earth's magnetic field.

Particularly exciting to those scientists interested in the composition of the lunar surface was the return of gamma ray data by Luna 10, providing for the first time specific compositional information. The results have been reported in some detail by Vinogradov et al. (1966). The instrumentation flown consisted of a scintillation counter with a 30×40 mm sodium iodide crystal and a 32 channel analyzer. One of the major handicaps of the experimental arrangement was that the detector was carried within the spacecraft leading to excessive backgrounds.

The published conclusions state that the general gamma ray flux on the Moon's surface in the areas studied is of the order of 20 to 30 microroentgens per hour. Of this radiation about 90% is cosmic in origin due to cosmic ray activation. The cosmogenic isotopes that were identified were oxygen, magnesium and silicon. Although the early reports stated that there were no noticeable differences in the intensity of gamma radiation over the mare and terra areas, later reports indicated that the mare areas had a higher integrated intensity. 10% of the gamma ray flux was attributed to *K*, *U* and *Th*. Because of the low observed intensities from the nuclides the Russians concluded that the mare material was similar to terrestrial basalts and the highland areas sampled closer to terrestrial ultrabasic rocks. However, because of the high backgrounds the Russian scientists admit that the interpretations are preliminary and must be considered as such pending further and more optimal measurements.

In 1966 the United States also succeeded in soft landing a Surveyor 1 on the lunar surface on Oceanus Procellarum. The spacecraft carried little scientific instrumentation other than a panoramic TV camera and strain guages on the landing members. It performed remarkably well returning over 11000 pictures, part of which were transmitted on the second lunar day after the spacecraft survived the lunar night. A clearly evident observation was that the mare material in the vicinity of the Surveyor was fragmental, poorly sorted, with low cohesion but substantial bearing strength. Based on the returns from the strain guage sensors, estimates of the dynamic bearing strength was about 6 to 10 psi.

The early lunar surface landings have since been followed by a number of other successful soft landings in various areas of the Moon. These include the USSR Luna 13 and the US Surveyors 3, 5, 6 and 7.

Surveyor 3 to Oceanus Procellarum was particularly noteworthy in that it carried a mechanical scoop or surface sampler. The design and use of this device has been described in detail by Scott and Roberson (1968). This sampler was used with extraordinary success, digging four trenches, performing six static bearing strength tests and seventeen impact or dynamic bearing strength tests. In addition the sampler was used to transport lunar material from one point to another. In one instance the lunar surface material was deposited on one of the landing pads for comparison with a color coded disk in order to determine the color of the lunar material.

With Surveyor 5, launched in the latter part of 1967, began a series of flights carrying to the lunar surface an alpha scatter experiment for analyzing the surface chemical composition, and a magnet assembly to determine the amount of material with high magnetic permeability at the landing site. Surveyor 5 was landed on the southwest portion of Mare Tranquillitatis, touching down inside a small crater about 9×12 m in horizontal extent and approximately 1.3 m deep. After landing the Surveyor vehicle slid downslope a short distance causing the footpads to produce shallow trenches in the lunar soil. One of these trenches was about 1 m long and 3 cm deep making it possible to draw some conclusions about the surface albedo. The material beneath the surface was quite surprisingly darker. This has been borne out in subsequent observations. The scientific results of the Surveyor 5 flight have been reported in considerable length in the Jet Propulsion publication, Tech. Rept. 32-1246, part 2 (science results). These may be summarized as follows:

The chemical composition determined by the alpha scatter experiment shows that the three most abundant elements on the lunar surface in order of decreasing abundances are similar to those on the Earth; oxygen, silicon and aluminum. The relative amounts are in fact similar to those of terrestrial basalts.

Observations of the magnet assembly by the onboard television camera showed considerable material of high magnetic susceptibility. The quantity of magnetic material adhering to the magnet was comparable to that expected from a pulverized basalt with 10 to 20% of magnetite and not more than 1% of admixed metallic iron. The estimated particle size attracted by the magnet was less than 1 mm.

Estimates of the albedo of the undisturbed parts of the moon near the Suveyor gave a figure of 7.9 plus or minus 1%. Debris ejected on the lunar surface near the footpads were estimate at 7.5 plus or minus 1%.

An interesting experiment was performed which involved firing the Surveyors vernier engine against the lunar surface. The observed erosion was attributed to the removal of particles by exhaust gases blowing along the surface and the explosive blowout of entrapped gas and soil from directly beneath the nozzel immediately after engine shutdown. Analysis of the data indicates a surface permeability comparable to terrestrial silts.

Since the results of the alpha backscatter experiment are strongly suggestive of basaltic composition and the Surveyor 5 site appears typical of the mare areas, the conclusion is drawn that differentiation has occurred and is probably due to internal heat sources. This conclusion is considered as being consistent with the point of view of mare basins filled with basaltic volcanic flows.

In November of 1967, another Surveyor was landed in Sinus Medii. The landing area was flat, though near a mare ridge. Like the previous Surveyor 5 mission, a variety of measurements were made of the surface composition, surface bearing strength etc. The results of this mission are summarized in the JPL Tech. Rpt. 32-1262 (1968). The lunar maria are reported as being remarkably uniform chemically, topographically and in physical properties. The depth of the debris layer is directly related to crater density whereas the coarse fragments are inversly related. The results of the alpha scattering experiment are that the elemental composition of the lunar surface at the Sinus Medii site is quite similar to that found by Surveyor 5 in Mare Tranquillitatis. The results were considered as suggestive of the fact that the elements were present on the lunar mare surface in the form of oxides, making up minerals and compounds similar to those found under terrestrial conditions.

The final Surveyor 7, was launched in January of 1968, and soft landed in the highland of the Crater Tycho (11.37° W longitude and 40.87° S lattitude). The payload included a television camera, an alpha scattering experiment, a surface sampler instrument, 3 auxiliary mirrors for observing special areas of vernier-engine jet impingement on the surface and magnets attached to two footpads and installed in the door of the surface sampler scoop.

Some of the scientific results, observations and conclusions are listed below (see JPL Tech. Rept. 32-1264, 1968). The general impression is that one observes more large rocks, fewer craters and a thinner debris layer in the highland area of Tycho as compared to the mare areas of the earlier Surveyor fligths. This is consistent with a relatively younger surface. Some of the rock fragments were observed to be coarse grained and to show evidence of crystals. By using the surface sampler it was possible to analyze three different types of lunar samples: the undisturbed lunar surface, a small rock and a disturbed area with exposed subsurface material. Within experimental error, the composition of all three samples is similar to that of the mare material examined in the earlier Surveyor fligths. The only reported exception is that the iron group of elements (Ti to Ni) is significantly less in the highland samples in comparison to the mare material. The samples analyzed in the Surveyor 7 landing have been grossly characterized as basaltic with low iron content.

On the basis of the above observations and analysis, the feeling exists that there is strong circumstantial evidence for some melting and chemical differentiation of the surface. If the iron content is lower, as the alpha scatter measurements seem to indicate, then the highlands have a significantly lower rock density than that of the mare material.

Finally we come to the most recent and perhaps the most spectacular events of all, the Apollo 11 and 12 lunar missions and their great scientific implications. As this manuscript is being prepared, two successful Apollo landings have occurred. During July, 1969, the Apollo 11 astronauts landed in the southwestern part of Mare Tranquillitatis, approximately 10 km southwest of the crater Sabine D. The astronauts succeeded in returning some 22 kg of lunar soil and rocks.

The Apollo 12 landing in November, 1969, was at a site south-southwest of Copernicus in Oceanus Procellarum. This second mission was even more successful in terms of both sample collection and the emplacement of a series of geophysical experiments. The Apollo 11 samples have been thoroughly analyzed and the results reported in the literature. The Apollo 12 samples are now being distributed to the scientific community for detailed studies. Some of the preliminary results will be described in chapter 4.

Summary

The above sections have been a brief chronicle of the events and findings that have occurred in the program of space exploration during the last decade. Some of these efforts and results can only be described as spectacular. As so frequently happens in science the observations and findings have generated as many if not more questions than answers. The rest of this book shall be devoted to the efforts being made towards an understanding of the chemistry of the moon and planets.

References

Annon.: Highlights of Press Conference at the Moscow House of Scientists, Pravda, 1966.

Fielder, G., Wilson, L., Guest, J. E.: The Moon from Luna 9. Nature **209,** 851—853 (1966).

Gault, D. E., Quade, W. L., Oberbeck, V. R., Moore, H. J.: Luna 9 Photographs: Evidence for a Fragmental Surface Layer. Science **153,** 985 (1966).

Kuiper, G. P., Strom, R. G., Le Poole, R. S., Whitaker, E. A.: Russian Luna 9 Pictures: Provisional Analysis. Science **151,** 1561—1563 (1966).

Lipskiy, Y. N., Pskovskiy, Y. P., Gurshteyn, A. A., Shevchenko, V. V., Pospergelis, M. M.: Current Problems of Morphology of Moon's Surface. Kosmischeskiye Issledovania **4/6,** 912—922 (1966).

Scott, R. F., Roberson, F. I.: Soil Mechanics Surface Sampler: Lunar Surface Tests, Results and Analysis. Geophys. Res. **73/12,** 4045—4080 (1968).

Significant Achievements in Space Science 1965. NASA SP-136, Office of Technology and Utilization, Washington, D.C., 1966.

Walter, L. S., Lowman, P. D. jr.: Significant Acievements in Space Science. NASA SP-155, Office of Technology and Utilization, Washington, D.C., 1967.

Space Research, Directions for the Future. Space Science Board, National Academy of Sciences-National Research Council 1965.

Vinogradov, A. P., Surkov, I. A., Chernov, G. M., Kirnozov, F. F.: Measurements of Gamma Radiation of the Moon's surface by the Cosmic Station Luna 10. Geochemistry no. 8, pp. 891—899. V.I. Vernadsky Institute of Geochemistry and Analytical Chemistry, Moscow, USSR, 1966.

Chapter 2: Instruments Used for Compositional Exploration

We will now describe some of the accomplished orbital and surface experiments and the instrumentation used for the compositional analysis of the moon and planets. Devices such as the alpha-back scatter instrument, the surface gamma-gamma density measurement, and the remote orbital gamma ray spectroscopy experiment have been successfully flown. Others, such as the Luna 12 (Lunar X-rays and cosmic ray background experiment), have achieved modest success meeting their objectives only in a limited way.

Scientific Objectives

One of the primary objectives in a program of planetary exploration is the determination of chemical composition and its correspondence to the elemental distributions on the earth's surface. This question has been discussed by Konstantinov et al. (1968), who examined the general problem of what chemical criteria might enable investigators to make valid determinations based on terrestrial analogues about the types of rocks or soil present on the moon. Figure 2.1 summarizes the way various rock forming elements fall into groups for the common lithospheric terrestrial rocks, ranging from ultra-basic rocks like dunites to the highly acidic granites. Included also for comparison are sedimentary rocks and stony meteorites. The following conclusions, based on these groupings, have been drawn by Konstantinov et al.

1. Sodium content may be used to distinguish ultrabasic (0.57 %) and sedimentary rocks (0.66 %) from the remaining ones. It will be impossible to distinguish neutral (3 %) from acid rocks (2.77 %). The situation is somewhat more favorable with respect to the sodium contents to basic rocks (1.94 %) and neutral rocks (3 %).

2. Magnesium values enable easy determination of ultrabasic rocks because the concentration (about 25 %) is higher in ultrabasic rocks than in other rocks by large factors (like 6-46). Magnesium is also separated in concentration in other rocks, thus making it an important signature element.

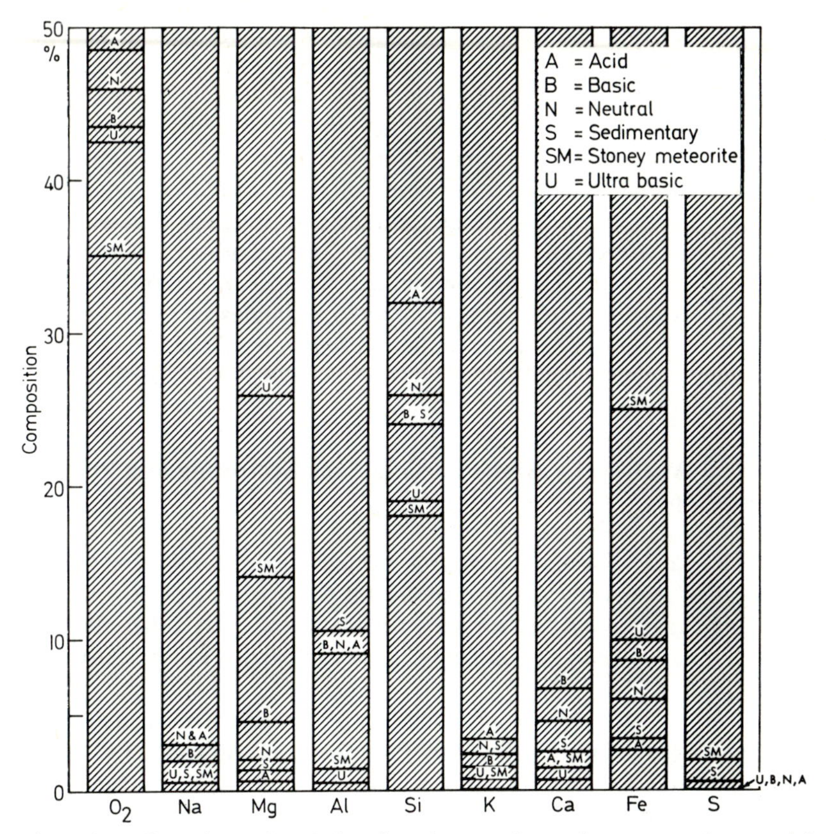

Figure 2.1 Groupings of rock forming elements for various types of terrestrial rocks

3. Aluminium is useful for classifing a rock as ultrabasic, because its concentration in ultrabasic rocks is quite low (0.45 %). The use of aluminium for distinguishing basic rocks (8.67 %) from neutral (8.85 %), acid (7.7 %), and even sedimentary (10.45 %), apears to be difficult.

4. Potassium content may be used to distinguish ultrabasic (0.3 %), basic (0.83 %), and other rock categories. Low calcium concentrations point to ultrabasic rocks (0.7 %). A significantly higher concentration (6.72 and 4.65 %) indicates basic or neutral types, while intermediate concentrations (2.53 and 1.58 %) relate the specimen to sedimentary or acid rocks.

5. Concentrations of sulfur and titanium vary considerably in different rocks, but usually are quite low except in the chondrites which have anomalously high values.

6. Iron content may be used as a criterion for ultrabasic rocks (9.58 and 8.56%), neutral (5.85%), or acid and sedimentary rocks (2.7 and 3.3%).

Konstantinov et al. conclude that the significant elements are magnesium, potassium, calcium, sulfur, and iron. We accept this conclusion with reservation. To use some of their criteria would require analyses to be performed with an accuracy and precision that would be difficult, if not impossible, to achieve under flight conditions. Their study does show, however, that by combining a number of determined parameters one can hope to obtain more valid estimates of rock types. An example of this will be given in our discussion of the alpha scatter experiment.

Chemical Analysis by Alpha Particle Back Scattering and Proton Spectroscopy

The alpha particle back scattering experiment carried for the first time on Surveyor V represents the first successful performance of in situ compositional analysis on the surface of a planetary body other than the earth. Similarly successful missions have been carried out on Surveyors VI and VII.

The use of scattered alpha particles for obtaining chemical analyses of surfaces was originally proposed by Professor S. K. Allison of the Enrico Fermi Institute for Nuclear Studies. The technique was investigated by A. Turkevich, and the results published in a paper "Chemical Analysis of Surfaces by the Use of Large Angle Scattering of Heavy Charged Particles" (1961). Turkevich presented experimental data showing that the energy spectra of back-scattered alpha particles were characteristic of the elements present in the scattering material. An outgrowth of these results was a proposal that a rugged, compact, analytical instrument could be designed and built for obtaining chemical composition of the lunar surface under the flight conditions being considered for the Surveyor Lunar Landing missions. Patterson et al. (1965) summarize the main features of their proposed method as follows:

The energy spectra of the large-angle, elastically scattered alpha particles are characteristic of the scattering nuclei. In addition, certain elements undergo alpha, proton (α, p) reactions, producing protons with characteristic energy spectra. By analyzing these energy spectra, it is possible to determine the chemical composition of the material bombarded by the alpha particles.

The method is described as having good resolution, with some exceptions, for many of the light elements expected in rocks; hydrogen, for instance, can only be inferred indirectly. Resolution, as we will see, becomes poorer with increasing atomic weight so that elements like iron, nickel and cobalt cannot be resolved. Sensitivities, however, are better for the heavy elements than for the lighter ones. The quoted sensitivity for elements above lithium is approximately 1 atomic percent.

Principles

On the basis of principles of energy and momentum conservation, Rutherford, Chadwick, and Ellis (1930) showed that scattered energetic alpha particles carry off some fraction of their original kinetic energies; the maximum fraction of the initial energy remaining is given by the following expression:

$$\frac{T_{max}}{T_0} = \frac{\left(\frac{4\cos\theta}{A} + \sqrt{1 - \frac{16}{A^2}\sin^2\theta}\right)^2}{\left(1 + \frac{4}{A}\right)^2},$$

where θ = scattering angle, A = mass number of the scatterer, and T_{max}/T_0 = fraction of kinetic energy remaining in the scattered alpha particle. For scattering angles close to 180 degrees (back-scatter direction) the above expression reduces to:

$$\left(\frac{T_{max}}{T_0}\right)_{180^\circ} = \left(\frac{A-4}{A+4}\right)^2.$$

T_{max}/T_0 represents the high energy cutoff or the minimum energy loss associated with a single backscatter collision.

Figure 2.2 shows some typical alpha-scattering spectra from thick targets of some pure elements obtained by Patterson et al. (1965), with the type of apparatus shown in Figure 2.3. It can be seen from Figure 2.2 that the spectra from the pure elements are of two types. For high Z elements (greater than about 20), the spectra characteristically are nearly rectangular with a distinct high energy cutoff. The spectra of the low Z elements such as carbon, magnesium, and aluminium show structure; the high energy endpoint is still present, but it is harder to identify.

Figure 2.4 shows the regularity in the relationship between the end-point energy fraction and the mass number of the scattering element. The observed data points are in excellent agreement with the mathematical expression written above. The endpoint dispersions are seen to be large for the low atomic numbers, and considerably compressed for the high atomic numbers. Thus, the technique is most effective for the low

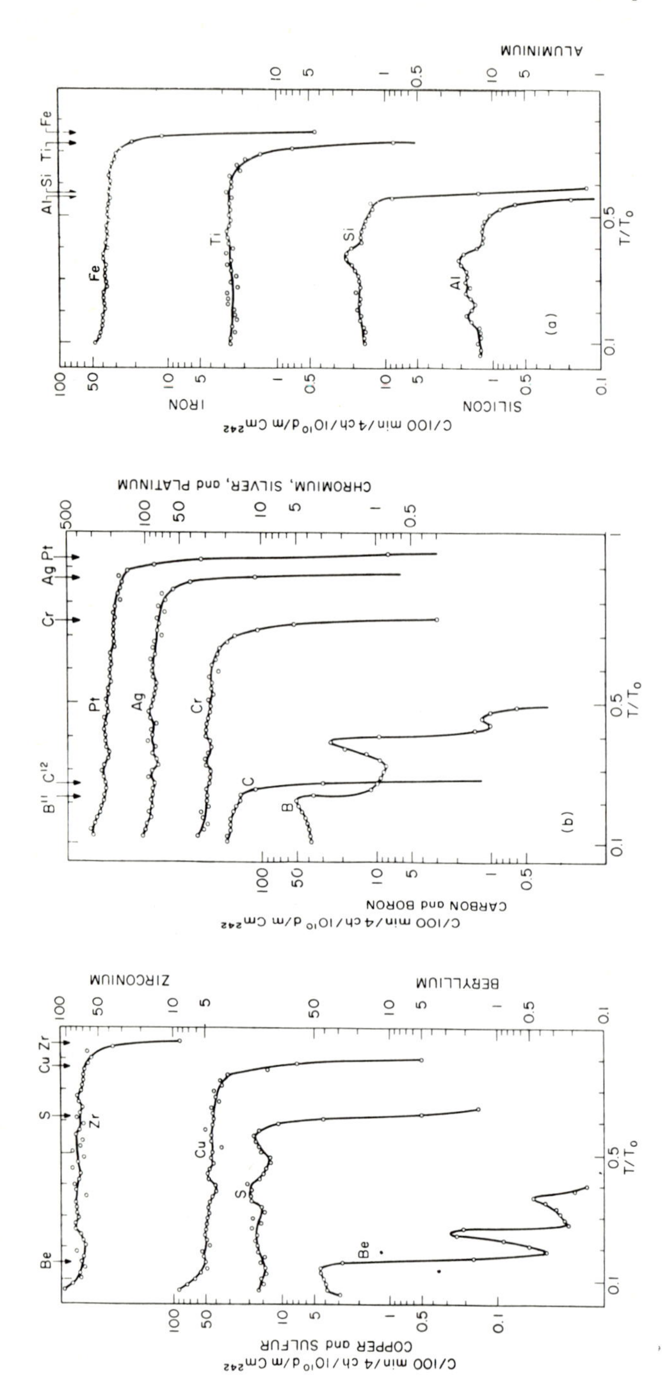

Figure 2.2 Typical alpha-scattering spectra of pure elements. (After Patterson et al., 1965)

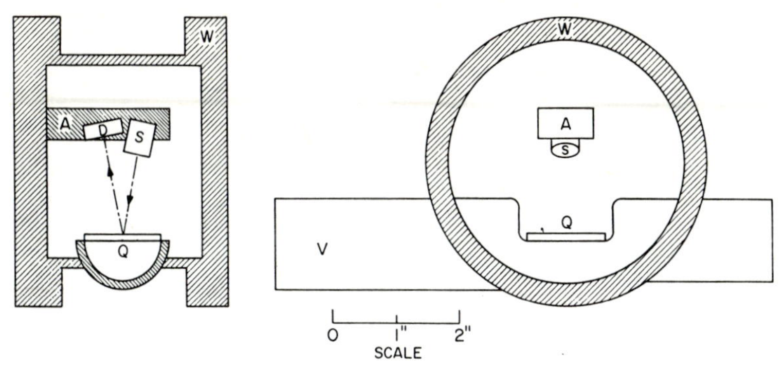

Figure 2.3 Type of apparatus used to obtain typical alpha-scattering spectra of pure elements

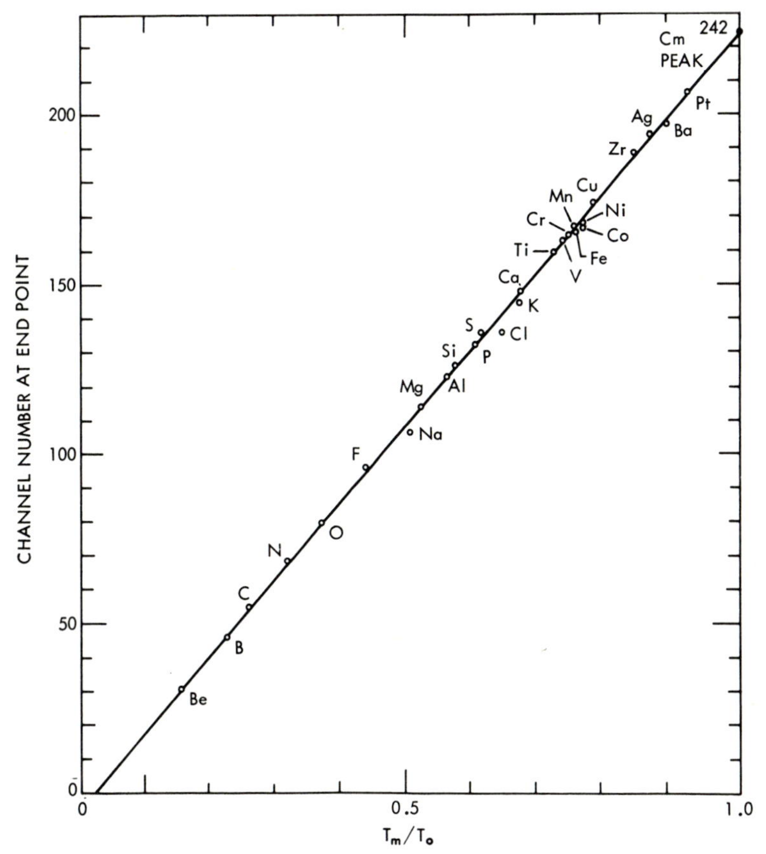

Figure 2.4 Agreement of observed and theoretical endpoints of the scattering spectrums of various elements as obtained with the research apparatus. The straight line, drawn through $T_{max}/T_0 = 1$ at channel 224 (the channel of the peak energy of the incident a particles), indicates how well the experimental end points agree with the predictions of Equation 1. The negative intercept at $T_{max}/T_0 = 0$ is consistent with the positive energy threshold of the analyzer. (After Patterson et al., 1965)

mass ranges. Since most instruments to date are said to have an overall resolution of about three percent, it becomes difficult to resolve elements with atomic numbers greater than 25. An additional complication is caused by the presence of several isotopes among the heavier elements, and the presence of several isotopes of the same mass number in different elements.

For the high Z elements, the shape of the spectra and the intensities depend on Rutherford scattering of the alpha particles. Given a particular geometry, one can, in principle, calculate intensities from theoretical coulomb-scattering cross sections and the energy loss characteristics of the scattering substances. In practice there are complications introduced by the incompletely known energy loss cross sections, and by the multiple scattering phenomena; both effects are most pronounced at low energies.

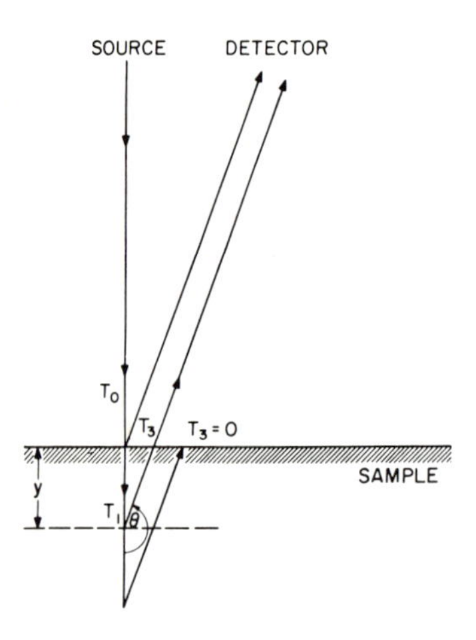

Figure 2.5 Geometrical relationships in α-particle scattering from a thick target. In this diagram the distances from the source and detector to the sample are assumed to be very large (centimeters) in comparison with the penetration of the α particles into the sample (microns). (After Patterson et al., 1965)

Following the description of Patterson et al., one can write the following expression for the scattered spectrum at an angle, θ, and as a function of the energy, T_3 (see Figure 2.5):

$$\frac{dI}{dT_3} = I_0 n \sigma(T_1) \frac{dy}{dT_3}.$$

We see that the intensity of the back scattered particles per steradian, at a given energy, T, per unit emergent energy, is a function of the following:

I_0 = intensity of the incident particles, n = the number of scattering nuclei/cm^3, $\sigma(T)$ = the alpha scattering cross section by the nuclei of interest at some emergent energy, T, and dy/dT_3 = variation in depth at which the scattering occurs with a change in emergent energy. The scattering cross section, in turn, is defined as:

$$\sigma(T_1) = \frac{5.184 \times 10^{-27} Z^2}{T_1^2(\text{MeV})} \cdot \frac{1}{\sin^4 \frac{\theta}{2}},$$

and thus depends on the atomic number, Z, of the scatterer, T_1—the energy of the alpha particle at the point of recoil, and on the scattering angle, θ.

It is obvious from the above expressions that one needs to know the nature of the energy loss by the alpha particle due to ionization, and to define more precisely the trajectories of the alpha particles in the target. Patterson et al. point out, in considering the spectral shapes, that the observed nature of the spectrum for high Z elements are in fact "consistent with Rutherford scattering cross sections and the stopping powers of the materials to the extent that they are known. Comparisons of the intensities of Rutherford scattering for various elements is best done near the high energy endpoints of the respective scattered spectra." In this region of the spectrum, it has been shown that the intensity of the alpha particles entering the detector (for the back scatter direction) has a $Z^{3/2}$ dependence. For elements below sodium, however, the regularity disappears and the scattering intensity becomes much higher than that predictable due to Rutherford scattering. Furthermore, the scattering varies in an irregular way from element to element.

The structure observed in the spectra of the low Z number elements is attributed to the departure from the $Z^{3/2}$ law. This has been related to the nuclear interactions with the target nuclei.

Proton Spectra from (α, p) Reactions

In addition to alpha scattering, one can also expect to observe characteristic proton energy spectra as a result of (α, p) reactions with a number of elements. For example, B, N, F, Na, Mg, Al, and Si yield protons with useful energies for measurement consistent with known nuclear masses and energy levels; the energy spectra and proton yields depend on the cross sections for this type of reaction. However, such reactions for incident alpha energies up to 6 MeV are impossible for nuclides such as ^{10}Be, ^{12}C, ^{13}C, or ^{16}O, ^{17}O and ^{18}O because of energy considerations.

For elements with nuclear charges above 20, the coulomb barrier becomes great enough to lower the (α, p) cross sections to very low values.

In comparison to the alpha scatter method, the proton intensities are often much lower than that of the scattered alpha particles. Incorporation of a detection system sensitive only to protons can be used to enhance the sensitivity of the overall alpha particle analytical system. It has been observed in rocks where, for instance, there is a relatively low abundance of aluminum and sodium relative to silicon and magnesium that the scattered alpha particle spectra are insensitive to variations in aluminum and sodium concentrations. However the proton spectra from aluminum and sodium, because they extend to higher energies and have characteristic shapes, are a much better basis for differentiation.

It must also be noted that the protons emitted may actually complicate the alpha scattering spectrum. This follows from protons depositing their energies in the alpha particle detectors, a process complicated by the fact that protons with their relatively longer ranges compared to the alphas may actually exceed the sensitive depth of the semi-conductor detectors biased specifically for alpha particle detection.

Data Interpretation

The methods for abstracting both qualitative and quantitative data from the observed spectra will be described in detail in the following chapter. In brief, the procedure involves the use of a library of elemental spectra and a least-squares computational technique, beginning with corrections made for background and instrumental drift (see Patterson et al., 1965). In order to obtain chemical analysis from complex spectra, the basic assumption is made that the atomic ionization energy loss of alpha particles in different elements varies as the square root of the mass number.

Instrumentation

The alpha scattering instruments flown on Surveyors V, VI, and VII, have been described by Turkevich et al. (1968). The equipment includes a sensor head deployable to the lunar surface, a digital electronics package kept in a thermally controlled compartment on the spacecraft, a deployment mechanism, and a standard sample assembly. The total weight of the assembly, including mechanical and electrical spacecraft interface substructures and cabling, is 13 kg. Power dissipation is of the order of 2 watts, increasing to 17 watts during times of heater use.

Figure 2.6 is a diagrammatic view of the alpha scattering head showing the internal configuration in cutaway. Figure 2.7 is a photograph showing the bottom view of the sensor head. The apparatus is a box approximately $17 \times 16.5 \times 13$ cm in dimension. The bottom of the box is a 30.5 cm

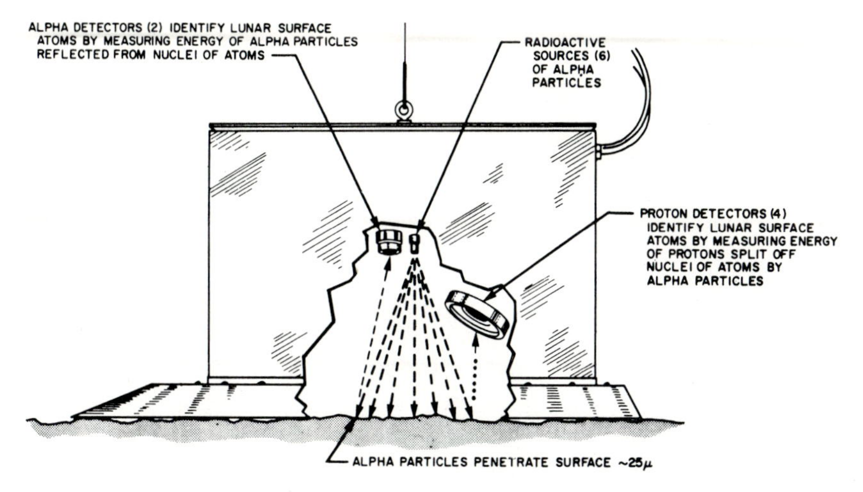

Figure 2.6 Diagrammatic view of alpha-scattering sensor head internal configuration

diameter plate to keep the box from sinking into the lunar material should it prove to be soft. The bottom of the sensor head has a sample port approximately 11 cm diameter. Six ^{242}Cm alpha sources are recessed 7 cm above the opening. The alpha sources are collimated in such a manner as to direct the emitted alpha particles to the surface material through the opening in the sensor head. Each collimator has a thin aluminum oxide film to prevent alpha particles, recoiling from the sources themselves, from reaching the sample area. The two alpha particle detectors are arranged to detect the back scattered alpha particles at an angle of about 174 degrees from the original direction. The alpha detectors are silicon, surface barrier types, 0.2 cm^2, with an evaporated gold front surface. In addition, thin films are mounted on the collimation masks to protect the detectors from both excessive light and alpha contamination.

The four proton detectors can also be seen in the bottom view of the sensor head. These are lithium drifted silicon detectors, 1 cm^2 each. These detectors are protected from the alpha particles by 11 micron thick gold foil. The proton detectors are backed by guard detectors in anti-coincidence mode so that only those events associated with the sample are counted. Solar events registering in the guard and the primary detectors are rejected by the electronics.

The outputs of both types of detectors are amplified and converted to a time-analog signal of a duration proportional to the energy deposited in the detector. Each type of detector works into a separate 128 channel analyzer. In practice, the outputs of the two alpha detectors were com-

Figure 2.7 Bottom view of the alpha backscatter head

bined before conversion. A separate mixer was used for the four proton detectors. A ratemeter was used to measure the frequency of events occurring in the guard detectors, but no information was obtained on the energy of such events.

Two techniques were used to provide calibration of the electronics: A pulser supplying electronic pulses of two known magnitudes at the detector stages of the alpha and proton systems on command from the earth; and small amounts of Einsteinium, an alpha emitter, located on the gold foil facing each proton detector and on the thin films located in front of the alpha detectors.

Figure 2.8 shows the deployment mechanism and sample mount. The deployment mechanism, in addition to being used for deploying the sensor head to the surface, is also used to stow the sensor during flight and to provide both background and calibration measurements prior to the surface measurements. As shown in Figure 2.8, there is a standard sample assembly covering the circular opening in the bottom of the sensor head during flight and during landing. This standard sample assembly serves both as a dust and light barrier, and also helps evaluate the instrument performance immediately after the lunar landing. Before deployment to the surface, the sensor head is held suspended about 50 cm above the lunar surface to obtain background data.

A typical operational sequence follows: 1. Measurements are first made in the stowed position. Data received from the standard sample and from the pulser calibration are compared with prelaunch measurements. 2. The supporting platform and standard sample are moved aside and the sensor head is suspended about 50 cm above the lunar surface in order to obtain background information on the detector response to cosmic rays, solar protons, and possible surface radioactivity. 3. Finally the sensor head is lowered to the lunar surface for the accumulation of surface data.

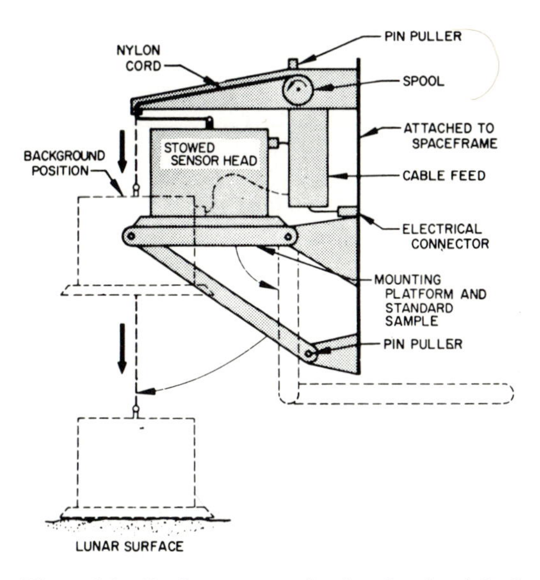

Figure 2.8 Deployment mechanism for the alpha backscatter experiment

Mission Description and Results (Surveyor V)

The Surveyor spacecraft landed in the southwest portion of Mare Tranquillitatis on September 11, 1967, at approximately 23 deg.E. latitude and 1.5 deg.N. longitude. The spacecraft landed on a slope of about 19.5 deg. on a crater wall. Table 2.1 summarizes the data accumulation times.

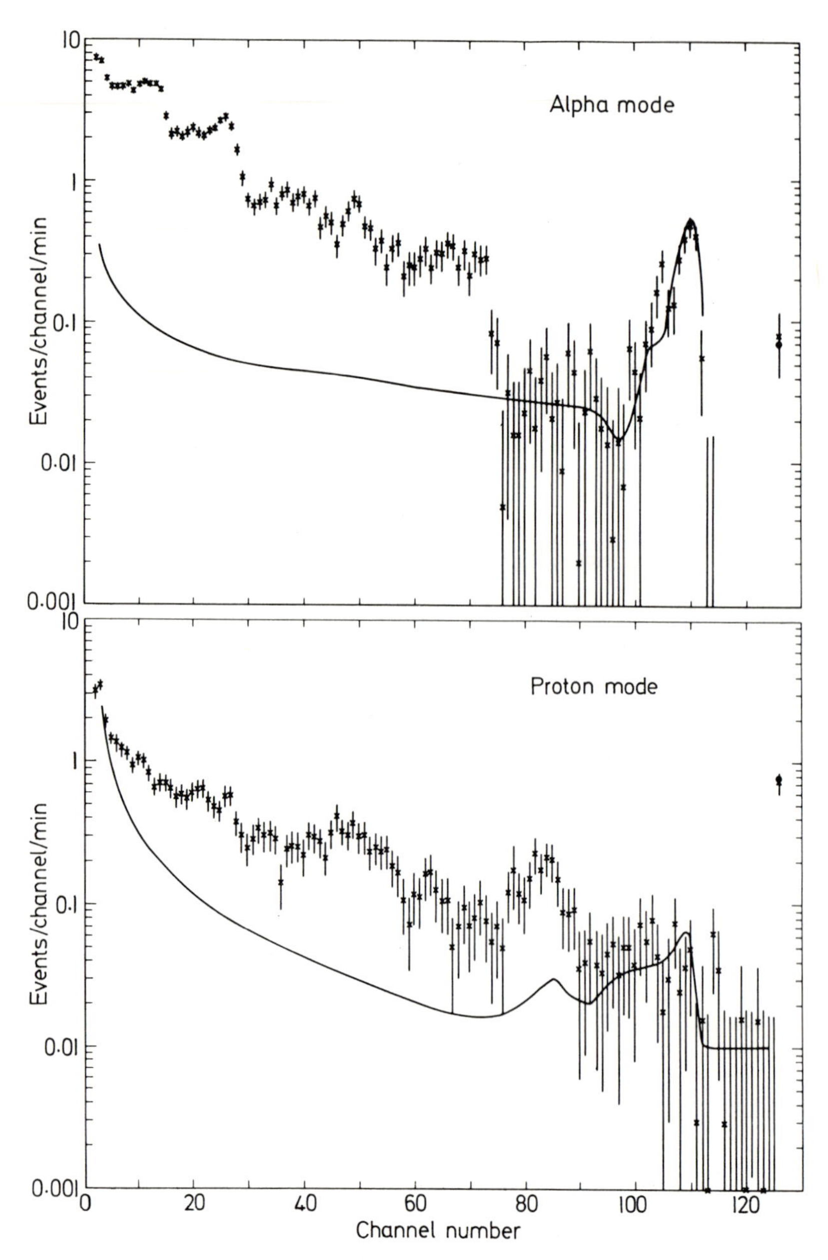

Figure 2.9 Measurement of standard glass sample on the moon after landing for a 60 minute interval. The error bars on the observed points (crosses) are for 1 σ statistical errors. The peaks observed in both modes are due to E_s^{254}. (After Turkevich et al., 1968)

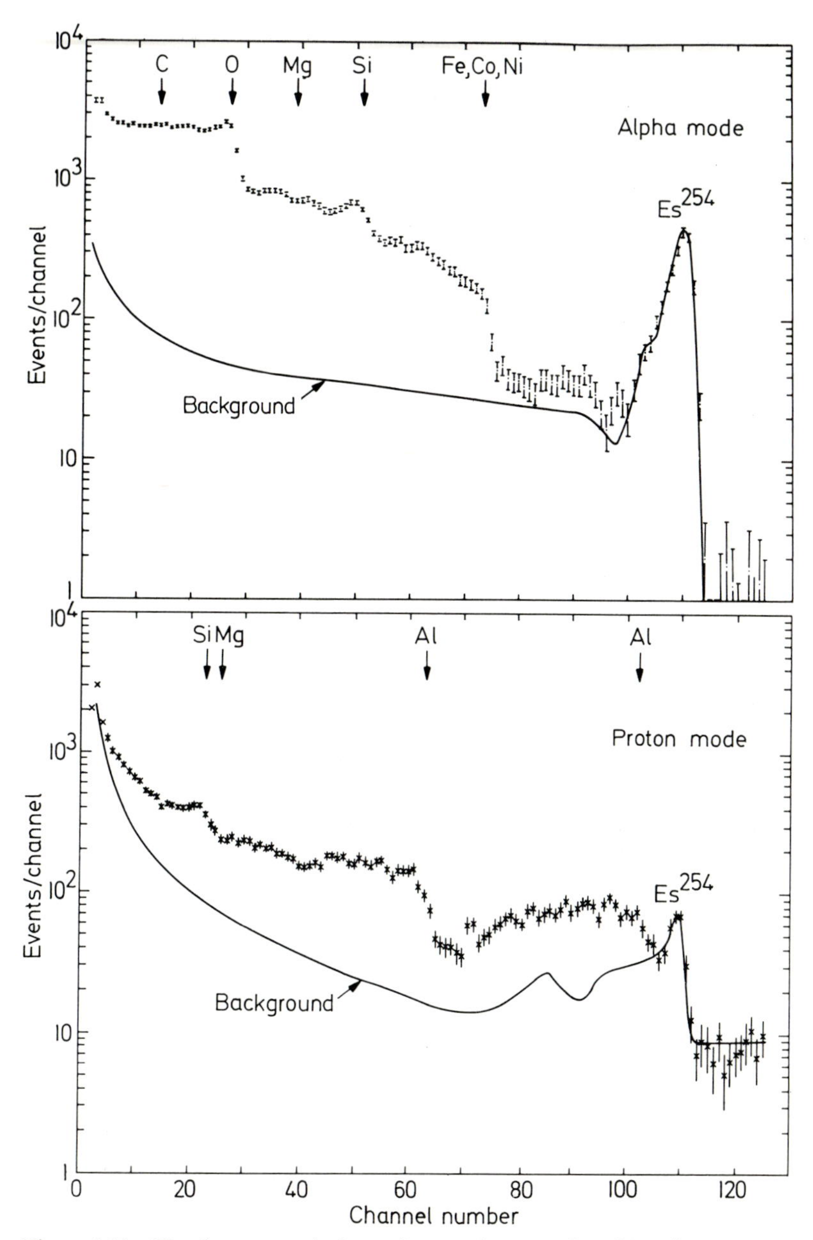

Figure 2.10 First lunar sample data taken on the moon for a 900 minute measurement time. The experimental points are crosses and the error bars 1 σ statistical errors. The peaks at approximately Channel 110 in both modes are due to E_s^{254}. Prominent features in some of the elemental spectra are shown by arrows. (After Turkevich et al., 1967)

Table 2.1. *Science data accumulation times*

Operational configuration	Accumulation time in minutes
Transit	20
Stowed (standard sample)	75
Background	170
Lunar surface sample 1	1056
Lunar surface sample 2	4005
Calibration	281
Total	5607 min (93.5 h)

(After Turkevich et al., 1967)

Two sets of measurements were made of the standard sample (an analyzed glass) referred to above. The first measurements were taken during transit to the moon, and the second set after touchdown on the lunar surface. The data shown in Figure 2.9 were taken on the moon over a 60 minute measurement period. The calculated results are shown in Table 2.2.

Table 2.2. *Analysis of standard glass sample on the moon*

Element	Atomic %	
	Surveyor V analysis*	Standard chemical analysis
Oxygen	56.4	58.6
Sodium	7.3	7.7
Magnesium	7.6	8.5
Aluminum	2.0	1.5
Silicon	20.2	17.2
Calcium	–1.5	–
Iron	8.1	6.5

* The Surveyor V results are normalized to 100% on a carbon-free basis. The glass was covered by a polypropylene grid that masked about 25% of the area. (After Turkevich et al., 1967).

The first lunar sample data taken in both the alpha and proton modes is shown in Figure 2.10. The total accumulation time was 900 minutes. The curves are plotted with the number of events per channel on a logarithmic scale as a function of channel number (in turn, a function of energy). The statistical errors are also indicated as a smoothed version of the background. Prominently shown are the peaks in both modes due to the Einsteinium at approximately channel 110. A number of distinct energy breaks are seen and identified by the arrows at the top of the diagram. Figure 2.11 demonstrates a computer analysis of the data from both modes of analysis. The observed lunar sample spectrum has been broken down into eight components: C, O, Na, Mg, Al, Si, "Ca", and "Fe". As pointed out by Turkevich et al. (1968), "Ca" represents elements of $28 < A \leqslant \sim 45$, and "Fe" represents elements $45 \leqslant A < 65$. The computer fit to the spectral data shown in Figure 2.11 was performed by first subtracting the background and possible heavy element contri-

Table 2.3. *Chemical composition of lunar surface of Surveyor V site*

Element	Atomic %*
Carbon	<3
Oxygen	58.0 ± 5
Sodium	<2
Magnesium	3.0 ± 3
Aluminum	6.5 ± 2
Silicon	18.5 ± 3
$28 < A < 65$**	13.0 ± 3
(Fe, Co, Ni)	>3
$65 < A$	<0.5

* Excluding hydrogen, lithium, and helium. These numbers have been normalized to approximately 100%.

** This group includes, for example, S, K, Ca, Fe. (After Turkevich et al., 1968).

Table 2.4. *Comparison of composition of solar atmosphere, the earth's crust and Surveyor V site.*

Element	Atomic %		
	Solar atmosphere*	Earth** Continental crust	Lunar mare Surveyor V site
O		62.6	58.0 ± 5
Na	1.2	2.6	<2
Mg	16.1	1.9	3.0 ± 3
Al	1.1	6.5	6.5 ± 2
Si	(20.0)	21.2	18.5 ± 3
$A > 28$	18.7	5.2	13.0 ± 3

* The solar values, from H.C. Urey. The abundances of only the heavy non-volatile elements are given, normalized to 20% for Si.

** The values for the crust of the earth are average values for the continents, from Howard J. Sanders, Chem. and Eng. News, Oct. 2, 1967. (After Turkevich et al., 1968).

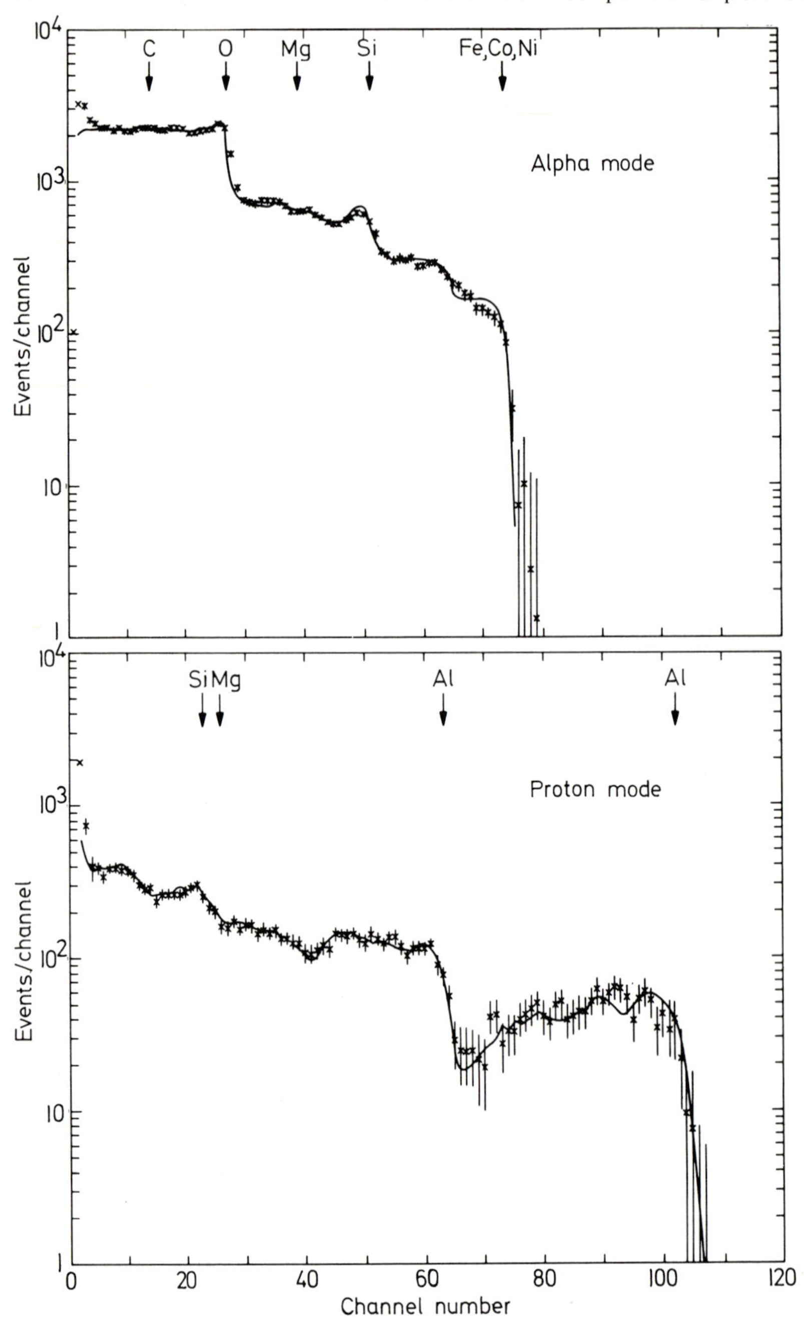

Figure 2.11 Computer analysis of first lunar sample. The smooth curves are calculated spectra based on an eight element library. (After Turkevich et al., 1967)

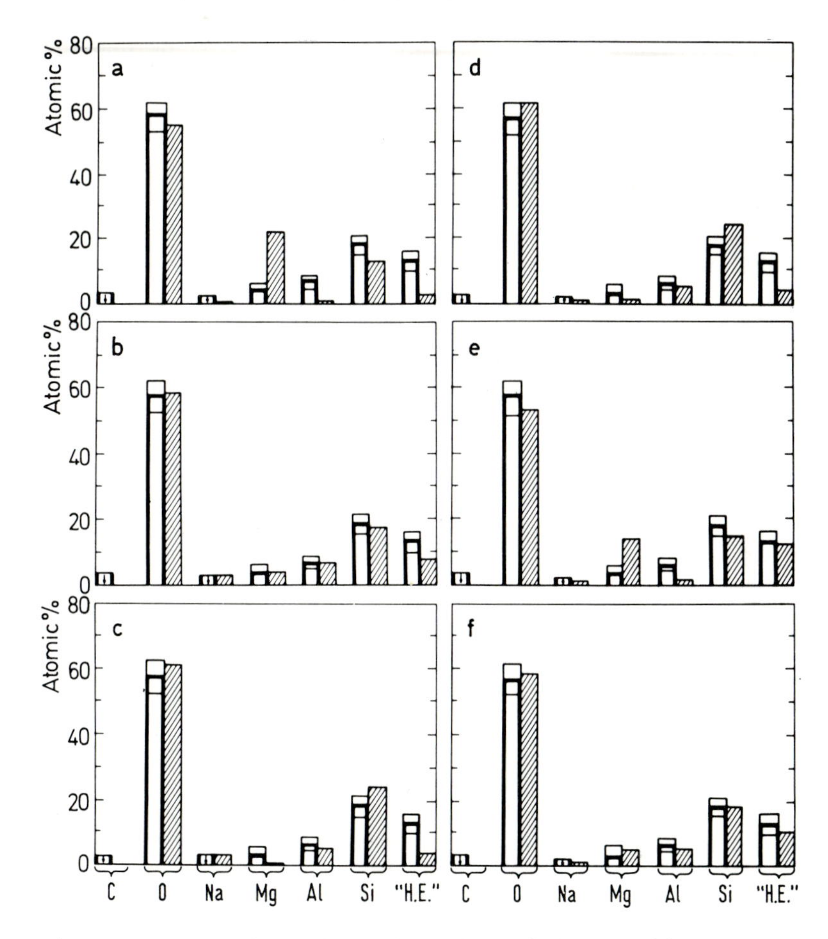

Figure 2.12 Comparison of the observed chemical composition of the lunar sample (open bars) with the average composition of selected terrestrial rocks such as: (a) dunite, (b) basalts, (c) granites, (d) tektite (Indo-Malayan body), (e) low iron chondritic meteorite, (f) basaltic achondrite meteorite. "H.E." are elements above Sc. (After Turkevich et al., 1968)

bution. The resolution of the spectra into only eight elements provides an excellent fit to the observations with very few systematic deviations. The one region of poor fit is between channels 63 and 74 in the alpha mode. Thus the elements in this region are reported as a composite.

Estimates of the chemical composition for the Surveyor V site are shown in Table 2.3 below.

These results have been further compared with other solar system bodies such as the sun and earth and are shown in Table 2.4.

From the above comparison, Turkevich et al. conclude that the Surveyor V results are more comparable to the chemical composition of the earth's continental crust than to the sun's outer regions. Thus, if the earth and moon were originally formed from solar type material, then the geochemical changes undergone by the material at the Surveyor V landing site and by the materials of the terrestrial continents must have been similar.

Finally in Figure 2.12 we see a comparison of the results obtained with various types of terrestrial rocks such as a dunite, basalt, granite, tektites, low iron chondrites, and a basaltic achondrite. On the basis of such comparisons, the conclusion is drawn that the Surveyor V site material falls most closely in the class of basaltic achondrites or terrestrial basalts.

Surveyor VI

The alpha scattering experiment flown on Surveyor VI was a duplicate of the one flown on Surveyor V. This has been reported by Turkevich et al. in *J.P.L. Technical Report* 32-1262 (1968). Once again a flat area in the equatorial zone was chosen as a landing site for the spacecraft. In this instance the landing site was Sinus Medii, and a landing was accomplished on Nov. 10, 1967.

Except for some minor problems, the equipment performed well (it was found necessary to turn off one proton detector which proved to be noisy).

A variant in the Surveyor VI mission involved the firing of the vernier engines, after about one week, in order to examine the effects on the surface in the vicinity of the spacecraft. This resulted in the vehicle moving about 2.4 m from the original site. One immediate consequence was that the alpha scatter sensor turned upside down. Although the instrument continued to function electronically, the indications were that there was now ^{242}Cm contamination in the sensor head caused by the possible rupturing of one of the source protective films. There was a large observed increase in the event rate of the proton system. Because of the opportunity offered for obtaining information on solar protons and cosmic rays, measurements were continued periodically for the remainder of the lunar day.

Results and Discussion of the Surveyor VI Alpha-scatter Experiment

Figure 2.13 shows the observed data for both the alpha and proton modes, and Figure 2.14 shows the computer analysis performed in the same

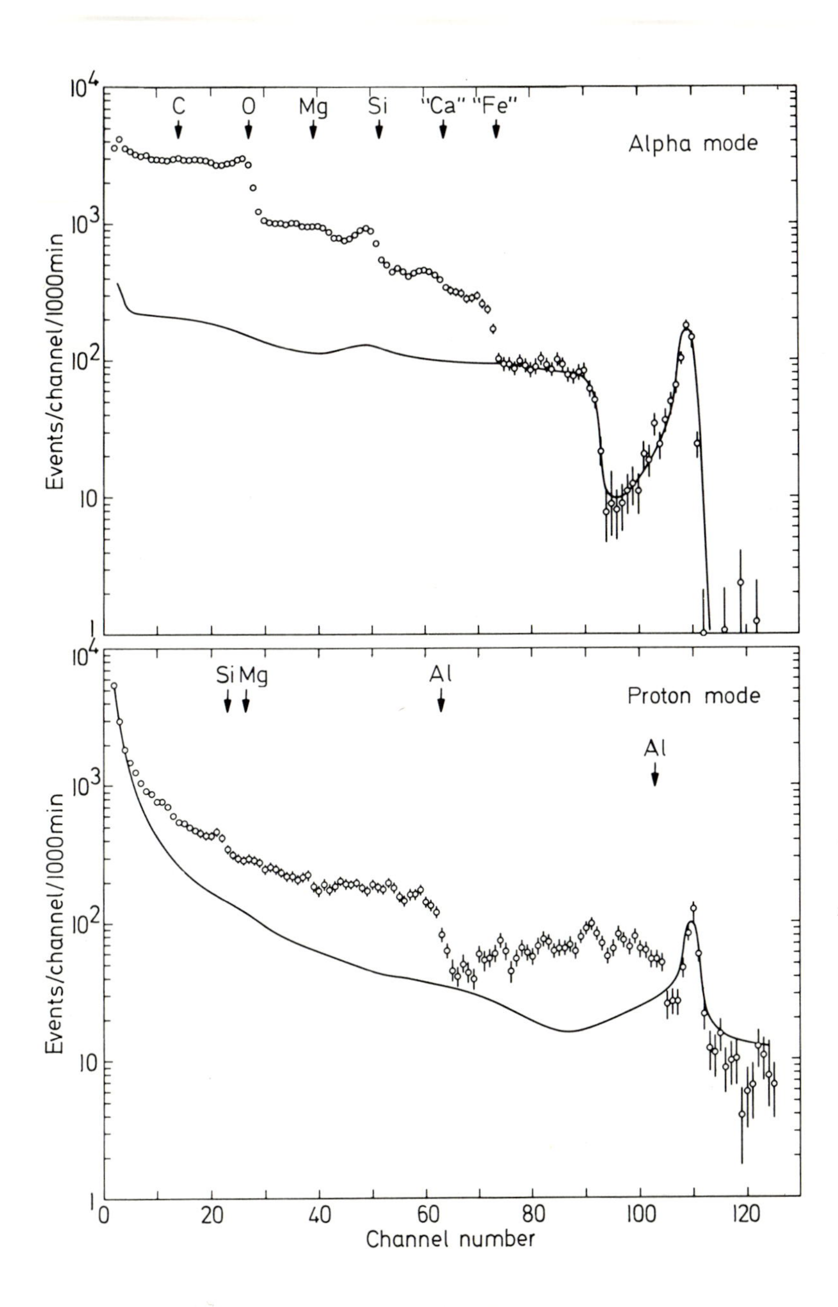

Figure 2.13 Surveyor VI lunar sample data. (After Turkevich et al., 1967)

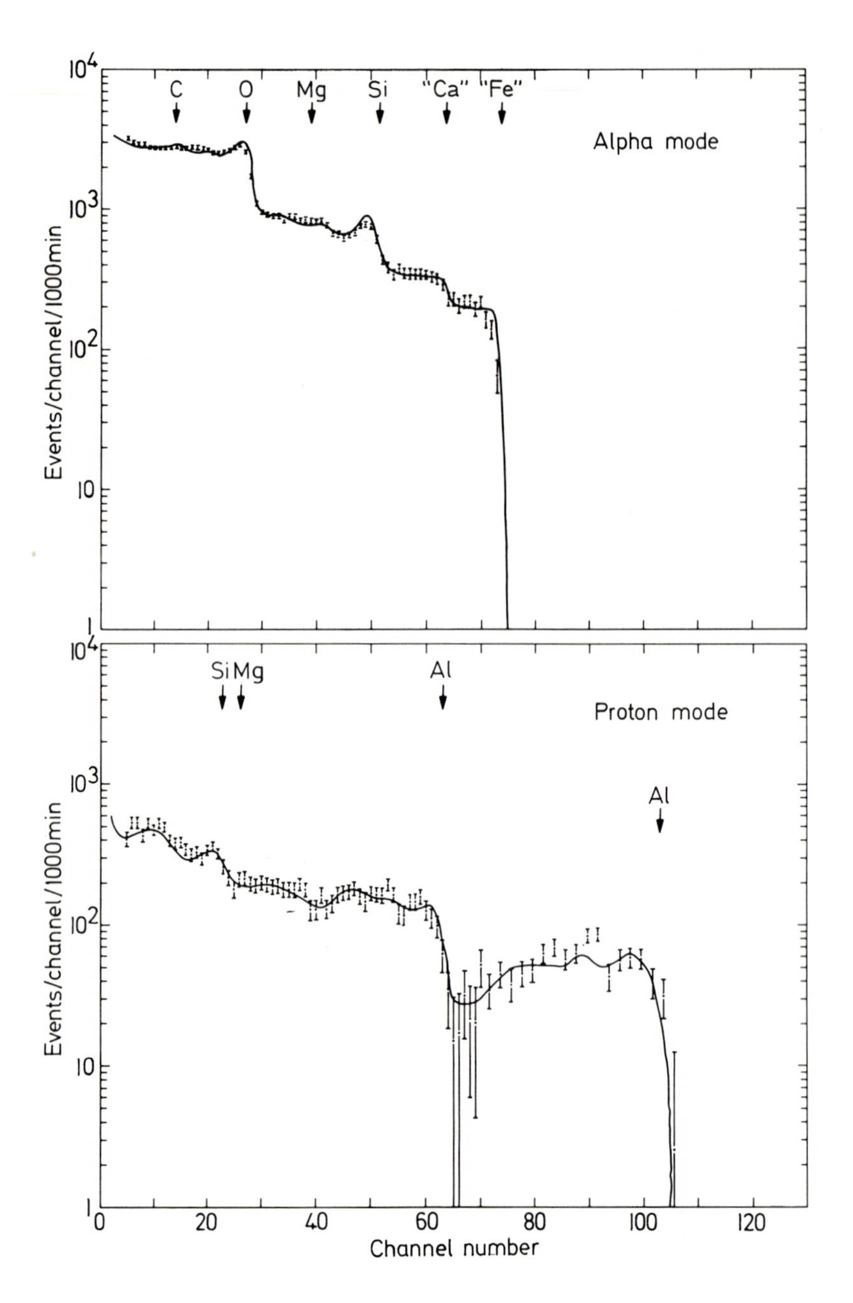

Figure 2.14 Computer analysis of Surveyor VI lunar sample data. (After Turkevich et al., 1968)

manner as in the Surveyor V measurements. An interesting comparison of the determined chemical composition is shown in Table 2.5. On the basis of these data, Turkevich et al. have stated in print that the chemical composition at the Surveyor VI site (Sinus Medii) is essentially the same as at the Surveyor V site. Thus, a considerable portion of the mare areas of the moon have a similar chemical composition. In fact, the lunar material is considered to be made up of chemical compounds and materials similar to those known on the earth.

Table 2.5. *Comparison of chemical composition at Surveyor V and VI sites*

	Atomic %*	
Element	Surveyor V**	Surveyor VI
C	<3	<2
O	58 ± 5	57 ± 5
Na	<2	<2
Mg	3 ± 3	3 ± 3
Al	6.5 ± 2	65 ± 2
Si	18.5 ± 3	22 ± 4
"Ca"***	13 ± 3**	6 ± 2
"Fe"****		5 ± 2

* Excluding elements lighter than beryllium.

** Surveyor V results were for the total of atoms heavier than silicon, a lower limit of 3% was set for "Fe".

*** "Ca" denotes elements with mass numbers between approximately 30 and 47 and includes for example P, S, K and Ca.

**** "Fe" denotes elements with mass numbers between approximately 47 and 65 and includes for example Fe, Ni and Co.

Surveyor VII (see Franzgrote et al., 1968)

The Surveyor VII mission was distinct from Surveyor V and VI in that the landing site selected was in a highland region near the rim of the Crater Tycho rather than in the equatorial mare region. One of the objectives of this, the last Surveyor attempt, was to sample material considered to be part of Tycho's ejecta blanket. The mission was also somewhat different in that the surface sampler (see Scott et al., 1968) provided a means of moving the alpha-scattering instrument from one position to another, thus making it possible to obtain data from three types of samples: undisturbed soil, small rock, and disturbed soil area near the nearby lunar surface.

As an interesting sidelight, the surface sampler played a key role when the deployment device failed to function properly. It was used to force the sensor head to the lunar surface and to provide the critically needed shade when the sensor head showed signs of exceeding the specified values of operational temperatures.

The alpha-scatter instrument flown was essentially the same as in the previous two flights except for a knob on top of the sensor head that served as a handle for the surface sampler. The preparation of the alpha sources was modified. The plates containing the curium were coated with carbon by vacuum evaporation in order to prevent aggregate recoil. This technique was partially successful. The alpha sources were also more intense than those flown previously, thus shorter accumulation times were required.

The Surveyor VII touchdown took place on Jan. 10, 1968, in an area less than one diameter north of the rim of Tycho. As mentioned above, difficulties were experienced in the deployment of the alpha-scatter experiment, but the problem was solved by using the surface sampler to help in the deployment of the sensor head. Ultimately, 3 different samples were measured. The first sample was undisturbed soil near the spacecraft. The second sample was an exposed rock which has been described as 5 by 7 cm in size, visible in the TV display prior to the beginning of the surface sampler operations. The third sample was in a trenched area, previously prepared during surface sampler manipulations.

An interesting observation reported by Franzgrote et al. is that the overall counting rate in the alpha mode for sample 2 (the rock) was about double that for sample 1. This data, and the television return, demonstrated that the rock sample was well centered in the sample area of the sensor head and, in fact, probably protruded slightly into the sensor head. By contrast, the intensity from sample 3, the trenched area, was lower than nominal, leading to the inference that the sample under examination was partially surface material and partially within the trench.

Results and Discussion

The Surveyor VII results have been discussed in the light of some of the instrumental problems that arose during the course of the mission. The program of chemical analysis was much broader and more productive than the previous missions because of the different samples examined. However, the experiments were somewhat limited because of factors such as delay in deployment and the circumstances leading to extended periods when the equipment was forced to operate at higher than prescribed

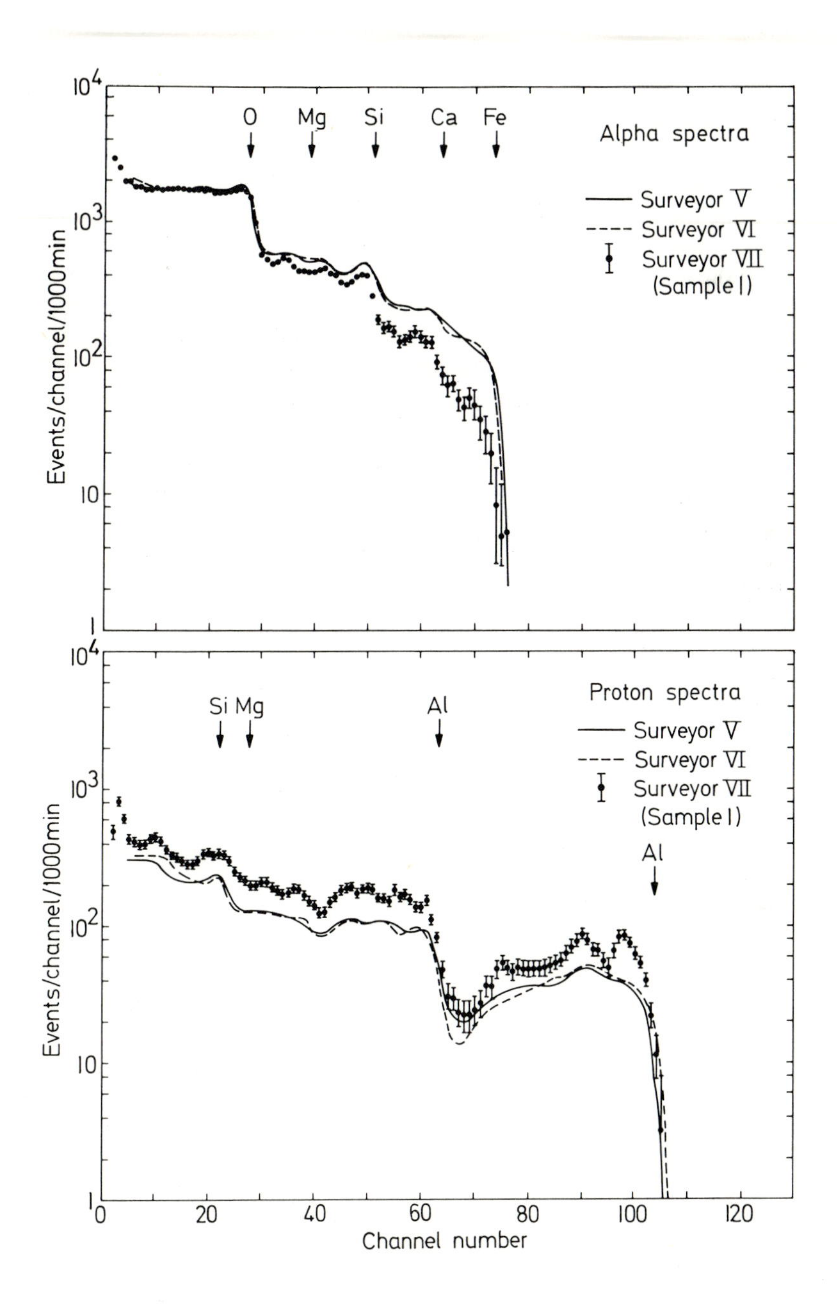

Figure 2.15 Comparison of Surveyor V, VI, and VII lunar sample data. (After Turkevich et al., 1967)

operating temperatures. Because high temperatures led to instrumental drift, proper data interpretation will require appropriate correction procedures to be made (if possible).

An interesting comparison of the spectra obtained for the three Surveyor missions is shown in Figure 2.15. The diagram includes only sample 1 for the Surveyor VII mission. In addition to background corrections, the curves have been normalized by multiplication factors that make the curves match in the oxygen region of the alpha mode. As reported, the only significant difference between the highland sample and the mare samples is in the lower content of the iron group of elements.

Table 2.6 shows a comparison of the chemical composition at the various Surveyor landing sites. The differences observed from the proton mode (Figure 2.15) are in magnitude rather than spectral shape, attributable tentatively to geometrical factors. There are as yet no firm results about samples 2 and 3 at this writing, although indicators are that the chemical composition of the rock (sample 2) is not materially different from the neighboring indisturbed lunar material (sample 1). The analysis of sample 3 will require supplemental laboratory simulations in order to determine the necessary geometrical corrections.

Table 2.6. *Chemical composition of the lunar surface at the Surveyor landing sites: Preliminary results*

Element	Chemical composition, atomic %*		
	Mare sites		Highland site
	Surveyor V**	Surveyor VI***	Surveyor VII
C	<3	<2	<2
O	58±5	57±5	58±5
Na	<2	<2	<3
Mg	3±3	3±3	4±3
Al	6.5±2	6.5±2	8±3
Si	18.5±3	22±4	18±4
"Ca"++	13±3+ (Ca + Fe)	6±2	6±2
"Fe"+++		5±2	2±1

* Excluding elements lighter than beryllium.

** Surveyor V results.

*** Surveyor VI results.

\+ Results from Surveyor V, in this case, included both the "Ca" and the "Fe" groups. A lower limit for "Fe" was set at 3%.

++ "Ca" here denotes elements with mass numbers between approximately 30 and 47 and includes, for example, P, S, K and Ca.

+++ "Fe" here denotes elements with mass numbers between approximately 47 and 65 and includes, for example, Cr, Fe, Co and Ni.

The inferences that have been drawn from the Surveyor VII analyses are as follows: 1. the lower content of "iron" group elements may be a contributing factor to the higher albedos of the highland regions; and 2. the lower "iron" group concentration, if truly characteristic of highland regions, can mean a bulk density of subsurface, highland rocks less than that of comparable material in the maria. If this is so, then the situation is analogous to terrestrial conditions where the material of the continental highlands is less dense than the basaltic ocean bottoms.

Lunar Surface Magnet Experiment

Beginning with Surveyor V, and continuing through Surveyor VII, a magnet assembly was installed on the Surveyor spacecraft, the objective being to determine the presence of material with highly magnetic permeability at the various landing sites. The principal investigator for these studies, J. Negus de Wys of the Jet Propulsion Laboratory, has described the experiments and results in detail in a series of reports (1967, 1968).

The basic premise was that materials such as free iron, magnetite, and Ni-Fe from meteorites, if present on the moon, would be attracted to a magnet mounted on the spacecraft close to the surface.

Magnet Assembly

The magnet assembly consisted of a magnetic bar of Alnico V and a non-magnetic control bar of Inconel X-750, both of the same dimensions ($5 \times 1.27 \times 0.32$ cm). The combined assembly was mounted on a footpad for easy television observation (see Figure 2.16). The magnetic axis of the magnetic bar was horizontal with the magnetic poles extending down the left and right sides of the magnet. A detailed magnetic strength plot of the face and poles of the magnet is shown in Figure 2.17. This plot is typical of the type of mapping done on each of the magnets flown. Readings were found to range from as high as 680 gauss on the pole faces, to zero down through the center of the bar. Calibration tests were also done on the magnetic field strength dropoff from the high areas on the magnets. Observations showed a strength of less than 0.38 gauss at a distance of 3.8 cm from the magnet. In laboratory simulations, powdered iron was collected from a distance of 1.93 cm by a 500 gauss magnet. It was also found that such a magnet would support about 20 g of magnetic material under terrestrial conditions.

Alnico V was chosen because of its high magnetic remanence at high temperatures, and because its magnetic properties seemed most suited for

Figure 2.16 Magnet assembly mounted on the footpad

lunar surface conditions. Since the material is brittle, both the magnetic bar and control bar were bonded to the bracket with RTV-60 (a bonding compound) in addition to screw attachments.

Laboratory Studies

A large variety of laboratory experiments were performed before and after flights in order to insure more accurate data interpretation of the data observed on the lunar surface. Some of these studies are described below:

1. Shadow progression studies: Since visual observations were a significant factor in the magnet experiment, the effect of shadow progression was studied using a full sized vehicle as well as a small model. Detailed studies were also performed on footpad 2 to determine the periods of optimum lighting.

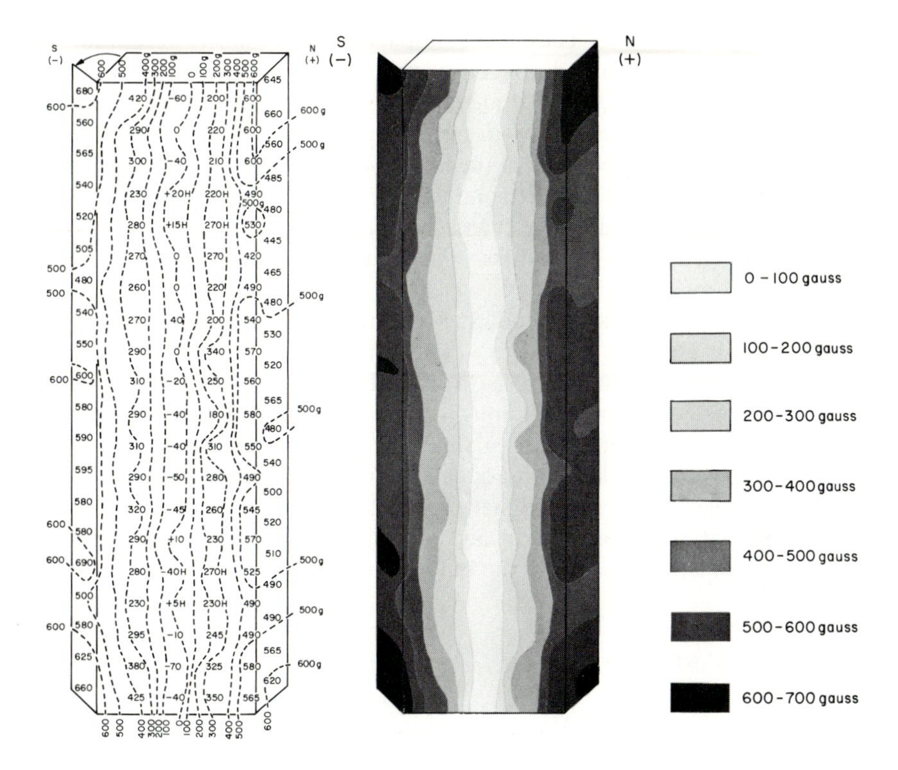

Figure 2.17 Magnetic strength plot

2. Impact tests with rock types: A number of different materials such as rhyolite, dacite, basalt, and peridotite were impacted against a footpad and magnet assembly under atmospheric pressure to determine the extent of "sticking" to the magnet assembly. Two different particle sizes (37 to 50 microns, 50 to 150 microns) were used in the studies. Particles were either driven against a stationary footpad-magnet assembly or the assembly was made to impact soil samples by plunging it into finely divided powder.

It was observed that only impact in basaltic powders caused appreciable adherence of the material to the magnet bar. Peridotite adhered to a much smaller extent, and acidic rocks hardly at all. This was attributed to the quantity of magnetite in the sample.

3. Impact tests with basalt containing added iron: Impact tests like those described in 2. on the preceding page, were performed using powdered basalts with varying amounts of pure powdered iron. The quantity of material adhering to the magnet increased with increasing iron content.

4. Vacuum studies: The effect of impact and jet exhausts were studied at a vacuum of 10^{-6} torr. Impact tests using basalts with added iron were repeated. In addition, an attitude control jet was fired into the samples at close range in order to set an upper limit to the amount of material which might adhere because of the firing of the Surveyor vernier engine. The firing jets caused very little adherence to the magnets. The vacuum impact tests produced a fine, even coating of material on the side of the bracket as well as the control bar. This material was easily removed at a return to atmospheric pressure.

5. Landing simulation studies: Six simulations of the landing mode were done in a trough filled with 37 to 50 micron basaltic particles. These landings were at about 1 m/sec from a height of approximately 1 m, penetrating to a depth of about 10 cm. The result was an adherent coating on the magnets, plus a fine film of adherent material on the control bar and bracket. Subsequently, the appearance observed in the trough as well as the test magnet proved to be similar to that observed in the actual Surveyor V flight and helped considerably in interpreting the vacuum adhesion seen in the lunar photographs.

Surveyor V Mission

The site of the Surveyor V mission has already been described under the alpha-scatter experiment. The behavior of the spacecraft on landing had some bearing on the magnet experiment. The spacecraft landed in a 9×12 m crater on a 19.5 degree inner north facing slope. Footpad 2 and 3 were on the downhill side as the vehicle performed a 1 m downhill slide forming a trench 3 to 10 cm deep. From this, and the lunar material observed on the front of the footpad, it was concluded that the magnet assembly contacted the lunar surface material. Calculations indicate that the magnet was exposed to about 500 cm^2 of the lunar surface.

Views were taken of the magnet assembly after touchdown, and before and after the vernier engine was fired. The vernier rocket engines were fired for 0.55 sec. An estimate of the dynamic pressure on the magnet assembly is given as 1 dyne/cm^2, sufficient to clean off the vacuum cohering material on the bracket and control bar, as well as that on the honeycomb structure and the zero gauss area down the center of the magnet.

Discussion of the Surveyor V Results

As reported by the principal investigator, Negus de Wys, the demands of the magnet experiment could not have been better executed. The requirement was for a solid impact with the lunar surface, and this was achieved (as examination of the pictures of the material covering the footpads showed). This was further confirmed by the nature of the trench produced by the downhill slide.

Interpretation of the lunar results is highly dependent on visual comparisons of the lunar photographs and those produced in the laboratory.

The following conclusions have been drawn from the various photographs studied:

1. Iron occurs on the lunar surface at the landing sites as magnetite, pure iron, and meteoritic Ni-Fe fragments.

2. Based on laboratory studies, an upper limit to the iron added to the naturally occurring rock is about 1% by volume of powdered iron.

3. Lunar magnet results agree with laboratory impact studies involving the 37 to 50 micron powdered basalt with no added iron.

4. The Surveyor V magnet results indicate less magnetic material than expected from entirely meteoritic pulverization and cratering of the lunar surface.

5. The magnet results support the reported findings of the Alpha-Scattering Experiment; that the observed lunar material has a composition similar to terrestrial basalts. This follows from the apparent agreement with the amount of magnetite to be expected in basaltic rocks.

Surveyor VI Magnet Experiment (see Negus de Wys, 1968)

The magnet assembly used in Surveyor VI was similar in construction to that flown on Surveyor V. Once again the assembly was mounted on footpad 2 to permit observation by the on-board TV camera.

Surveyor VI landed near a wrinkle ridge on Sinus Medii. In the initial landing, footpad 2 bounced about 25 cm and then slid about 6 cm. The magnet made no contact with the lunar surface material during this initial phase. After one week the attitude control jets were fired for 2.5 sec, causing Surveyor VI to hop from the original position to a position 2.4 m southwest of the original position. In this second landing, footpad 2 penetrated the lunar surface material approximately 10 cm and bounced, thus causing a horizontal displacement of approximately 12 cm. This mode caused the lower seven-eighths of the magnet to contact the lunar surface material. The strain gauge measurements made at the time indicated that the final small movement was a hop rather than slide, and resulted in an impact similar to those performed in the laboratory.

As in Surveyor V, data interpretation involved comparisons with the appearance of the magnet assembly after a variety of laboratory simulations of the type described for Surveyor V. These studies, and examination of the flight photographs, have led to the following conclusions (many of which are similar to the Surveyor V results).

1. A small amount of magnetic material is present in Sinus Medii. Based on laboratory studies, the amount of attracted magnet material appears to be less than 0.25% by volume.

2. The results of the magnet experiment appear to verify the basaltic composition found during the Surveyor V flight. The overall appearance is like that of Pisgah scoriaceous basalt.

3. There appears to be wider size distribution than previously indicated.

4. Extensive meteoritic addition to the lunar soil appears unlikely because of the low amount of magnetic material observed.

Surveyor VII (see Negus de Wys, 1968)

The magnet experiment on Surveyor VII was more elaborate than those carried on the previous 2 Surveyor missions. In addition to the usual magnet assemblies described above, and attached to the footpads (footpads 2 and 3 in Surveyor VII), two rectangular horseshoe magnets were also embedded in the door of the surface sample scoop (see Figure 2.18a and b). These magnets were $1.6 \times 0.96 \times 0.32$ cm, with magnetic flux strengths of about 700 gauss at the poles. The magnets were oriented so that the two south poles were next to each other in the center of the surface sampler door. These additional magnets were placed on the surface sampler in order to detect magnetic material at depth as well as on the surface. It was felt that the maneuverability of the surface sampler would also provide an effective means for testing any rocks located by the scoop for magnetic properties.

Observations

During the initial landing, footpads 2 and 3 penetrated the surface less than 6 cm; thus there was no contact between the footpad magnets and the lunar surface. The surface sampler magnets did however provide data from two locations. The sampler, in both instances, penetrated the fine grained homogeneous material to a depth of approximately 5 cm. A small amount of magnetic material was observed to outline the magnetic poles. After several trenching operations, the magnets showed more material adhering to the magnet poles. This was related to increased contact with the soil rather than to an increase of magnetic material with

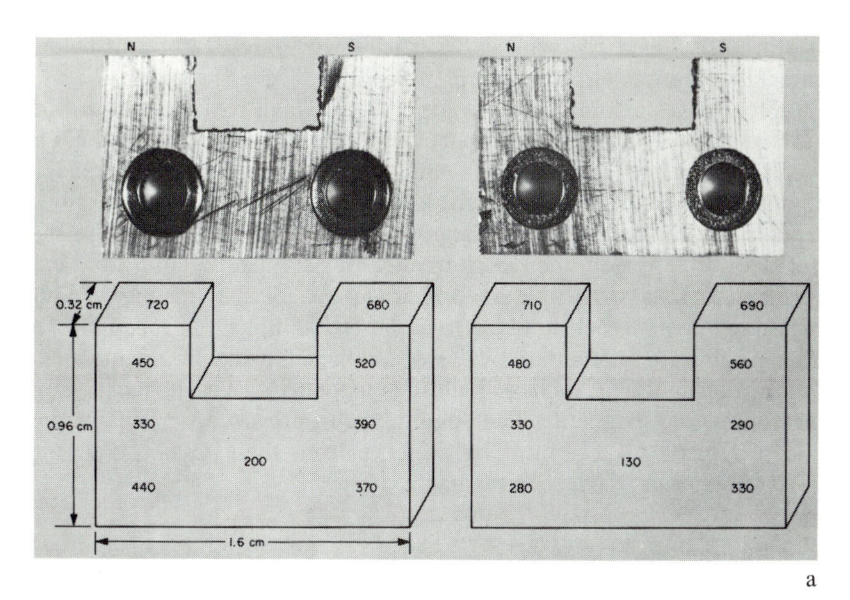

a

b

Figure 2.18 Magnet assemblies: (a) footpad, (b) scoop

depth. The size of the adhering particles was judged as being below camera resolution (approx. 1 mm).

Success was also achieved in attracting a small rock (1.2 cm) to the surface sampler magnets. This particular rock was observed in the TV display to have a lower albedo and a rounder and smoother shape than most of the other rocks in the area.

Because surface sampler magnets were used to a large extent in the Surveyor VII mission, their performance was studied by laboratory experiments similar to those performed on the footpad magnets. With regard to the small rock adhering to the surface sampler magnet described above, laboratory simulations were conducted using fragments of a number of materials such as pallasites, chondrites, basalt, nickel-iron meteorites, and magnetite. The only fragments attracted were magnetite and meteoritic nickel-iron. The evidence from these tests is that the lunar object must have been magnetic.

Conclusions

1. The appearance of the surface sampler magnets, following the bearing strength tests 1 and 2 on the lunar surface, is similar to those obtained in the laboratory using 37 to 50 micron powdered basalt (Little Lake Basalt). The resemblance was less when peridotite was used.

2. Particle size was less than 1 mm (camera resolution).

3. From the trenching operations, the material below the surface appears homogeneous and fine grained.

4. Comparisons of the Surveyor VII and the laboratory studies seem to give results compatible with basaltic powder with no observable addition of meteoritic nickel-iron (less than 0.25 % by volume).

5. The small amount of magnetic material in the upper 5 cm does not indicate an appreciable addition of meteoritic nickel-iron, or indicate churning of the upper layer by meteoritic bombardment.

There are some inconsistencies between the Surveyor VII alpha-scatter results and the magnet experiment, particularly with respect to iron content. The alpha-scatter experiment indicates iron concentration as being lower by a factor of 2 from the mare regions of Surveyor V and VI. By contrast, the magnet studies show no difference in concentration between the mare regions and Tycho. Some reasons for this have been advanced by Negus de Wys. The principal argument is that alpha-scatter experiment sampled only the top few microns of the surface which concievably could have been a mixture of several rock types. There were observations of lighter rocks of various sizes scattered about. The reader is referred to *JPL Technical Report* 32-1264 for more detailed arguments.

Density Measurements of the Lunar Surface By Gamma Ray Scattering

The work described below represents the first in situ measurement of the moon's surface density (A. A. Morozov et al., 1968). The experiment was flown on the USSR "Luna-13 Automatic Station" which landed in Oceanus Procellarum (18° 52′ N. latitude and 62° 03′ W. longitude) in December of 1966. The technique for determining the surface density was radiometric, depending on the relationship between surface density and the intensity of backscattered gamma radiation. The principles of the method are described below.

Gamma rays incident on material such as soil, for example, undergo three types of interaction. Two of these, photoelectric absorption and pair (electron-positron) production, result in the total disappearance of the incident gamma ray photon. The third process, Compton scattering, imparts an energy loss and direction change to the incident radiation.

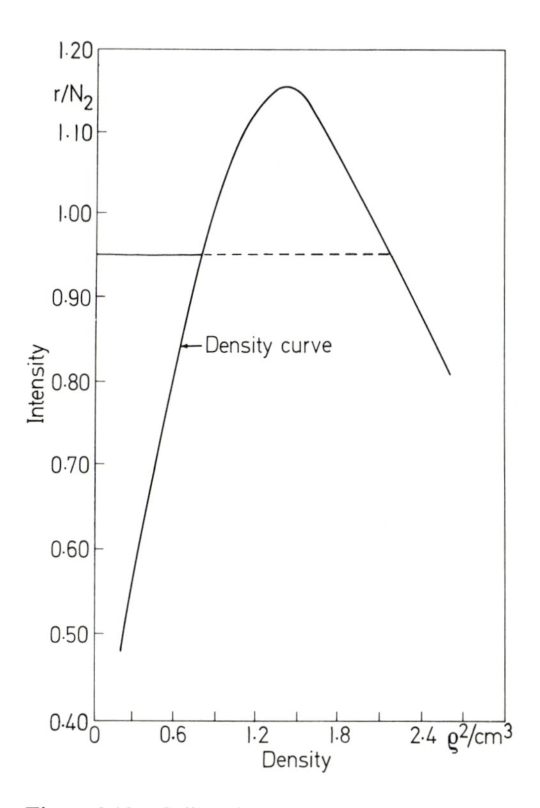

Figure 2.19 Calibration curve of the radiation densimeter

For radiation in the range of 0.5 to 3.0 MeV, the Compton effect is a fundamental one; the cross section for scattering is proportional to Z/A, where Z is the atomic number and A the atomic weight of the scattering element. It can easily be shown that the ratio of Z/A for all the elements commonly found in soil (except hydrogen) is 0.4 to 0.5. This is of considerable significance in the method being described because it demonstrates an independence of chemical composition.

The determination of density by gamma ray scattering depends on the functional relationship between the intensity of the scattered radiation and the density of the medium ($N=f(\rho)$). The type of relationship observed in practice is shown in a calibration curve in Figure 2.19. The curve is seen to have three features: an ascending portion, a maximum, and a descending portion. This is caused by two competing processes, scatter and absorption. The Compton scattering process is proportional to the number of electrons per unit volume. The intensity of the scattered gamma rays increases with density, as can be observed in the ascending portion of the curve at low densities. Beyond this point, absorption becomes more predominant and the curve falls. This behavior is a consequence of multiple gamma ray scattering leading to decreasing energies, and a resulting increase in absorption. The position of the maximum depends on the physical geometry of the apparatus. There is, for example, a relationship to the distance between the gamma ray source and the detectors.

The "radiation densimeter" flown on Luna 13 has been described as follows: The unit consists of a deployable sensor head and a monitor unit. A diagrammatic sketch of the sensor unit is shown in Figure 2.20. The sensor head contains the gamma ray source (Cs-137) as well as 3 groups of gamma ray detectors, each group in turn consisting of 5 counters. A lead shield is used to isolate the detectors from the source. The distance

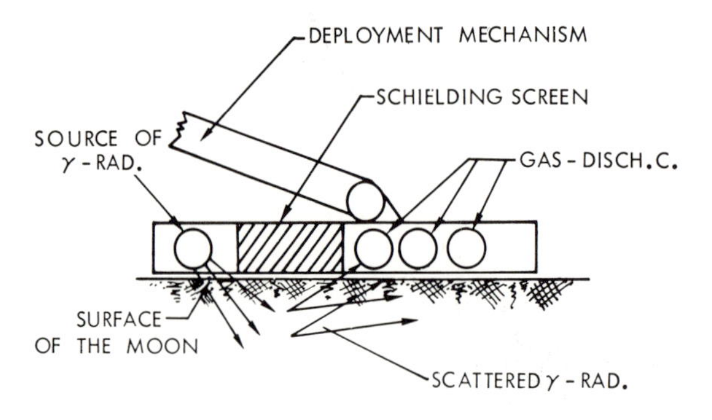

Figure 2.20 Radiation Densimeter

between the source and the three groups of detectors (base length) is different, and the device thus permits separate measurements for each of the three different base lengths.

The sensor unit flown was approximately $26 \times 5 \times 1$ cm, and had 2 lateral segments to ensure the proper orientation of the head on the moon's surface.

Measurements of the intensity of the scattered radiation were performed by counting pulses, using solid state counting circuits. The outputs of the three groups of detectors were separately integrated, and the resulting voltages were then transmitted. The individual groups of detectors were integrated by the telemetry at two minute intervals.

Data Interpretation

Data interpretation was done on the basis of ground calibration. A variety of materials of varying density, such as heavy concrete, light concrete, glass foam, etc. were used to determine performance. Calibration was performed over a range of densities from 0.16 to 2.6 g/cm^3. Simultaneously, information on the energy spectrum of the scattered rays was obtained. The plotted calibration curve is similar to the one shown in Figure 2.19.

In studying the instrumental performance, and in the preparation of calibration curves, the investigators paid special attention to the effects of surface roughness on the measured intensities. The results are shown in Table 2.7. It was observed that the effect of small craters or gaps depends on the density. For dense materials (the descending part of the curve), the existence of craters or loosely packed soil yields increased integral intensity measurements. By contrast, such pits or craters in the low density region lead to a decrease in counting rate. From an analysis of the data in Table 2.7, it has been concluded that small craters of the order of 3 to 4 cm, and at depths of about 2 cm, have only an insignificant effect on the variation of the counting rate for a given density. The errors are about 5 to 10%, depending on the position of the crater with respect to the sensor head. The effect of these craters also depends on the base length between the source and the detectors. In some instances the effects are in opposite directions for the small and long baselengths. Averaging the results can help to minimize the effects of uneven surface conditions.

Results

An actual density determination was performed during the Luna 13 flight. The measurements were reduced by reference to the calibration shown above (Figure 2.19). This curve was prepared relative to the intensities scattered by a material of density 2. Two possible values can be taken from the measurements: $\rho_1=0.8$, and $\rho_2=2.1\,g/cm^3$. The lower value is characteristic of light, foamy, and porous materials such as slag, pumice, glass foam, or light granular type materials. The upper value

Table 2.7. *Results of experiments on the determination of surface roughness effect on the counting rate of gamma-quanta by the radiation densimeter*

Form of surface's relief under the sensor	Counting rate	
	base l_{min}	base l_{max}
Even surface	1.0	1.0
Counters above recesses	0.986	1.08
Counters above recess; 1 cm gap under source	0.936	1.06
Counters above recess; additional gap under the shield.	0.958	1.10
Source above recess	0.974	1.00
Recess under the median part of the sensor	0.936	1.18
Recess under the source and the detectors	0.963	1.09
Sensor on "heaps"	0.804	1.21

(After A. A. Mozorov et al., 1968).

is associated with denser material such as rocks. The lower value is interpreted as being more in accord with other independent measurements made of the moon's surface (such as astronomical, photographic, and geophysical). Thus Mozorov et al. conclude that the figure $\rho_1 = 0.8$ is a more reliable estimate of the surface density down to a depth of about 15 cm (the range of effectiveness of this type of device).

Gamma Ray Spectroscopy from Lunar Orbit

The primary objective of the gamma ray experiments to is find out if the moon has undergone any substantial chemical differentiation during its development. This determination will be accomplished by the measurement and interpretation of gamma ray spectra emitted from the lunar surface. The major source of information is derived from gamma ray emission, from natural radioactive emitters such as ^{40}K, U, and Th, and from activity induced in the lunar surface by the incident cosmic ray flux.

The application of gamma ray spectroscopy to study lunar surface composition was proposed by Arnold (1958). An experiment was flown during a Ranger block 2 mission (Arnold, 1962; Van Dilla, 1962). Information was obtained about the isotropic flux of gamma rays in space, but lunar data was not obtained.

In 1966 the Russians reported a gamma ray experiment on the lunar orbiters Luna 10 and Luna 11 (Vinogradov, 1966). The experimental results were limited by the small crystal used in the detector, limited counting times, and an almost overwhelming interference by the spacecraft induced background. However, the experiment demonstrated conclusively that gamma ray spectra could be obtained from the lunar surface, and placed an upper limit on the contribution of naturally occuring radioactivity from broad areas of the moon.

We concentrate here on instrumental techniques, the results obtained from the Ranger and Luna missions.

Instrumentation

The instrument described below was used in the Ranger 3, 4 and 5 flights. A similar design was employed as a detector system aboard the Luna spacecrafts.

Figure 2.21 is a photograph of a full scale model of the Ranger 3, 4, and 5 spacecrafts showing the solar panels, the high gain antenna, the hexagonal control structure containing the electronic components, the spherical package housing the seismograph above its retrorocket, and the gamma ray detector in its spherical shell at the end of a six foot boom.

Figure 2.21 Full scale model of the Ranger spacecraft

A cesium iodide scintillation crystal was chosen at the time of the Ranger 3, 4, and 5 because: 1. the greater strength of the cesium iodide, and 2. the slower pulse-rise time makes the design of a phoswich discriminator easier. The lower light-emission of cesium iodide, as compared to an NaI(Tl) crystal, makes it less desirable in terms of energy resolution. Now, with improved electronics, photomultipliers, and, mechanical packaging techniques, NaI(Tl) detectors are being used. The size of the crystal used in the Ranger experiment was 2-3/4″ × 2-3/4″ right cylinder, and weighed 2-1/2 pounds.

The physical design of the detector is shown schematically in Figure 2.22. In assembly the CsI crystal is inserted in the plastic anti-coincidence mantel. The electronics circuits surrounding the photomultiplier tube include the phoswich rejection circuit (used as an active shield against cosmic radiation), and a unit-gain emitter-follower for transmitting the CsI signal down a six foot boom to the analyser.

The principal problem with the associated circuitry is that of distinguishing between the pulses arising in the plastic and those originating in the CsI over a wide dynamic range. Pulse shape rejection was accom-

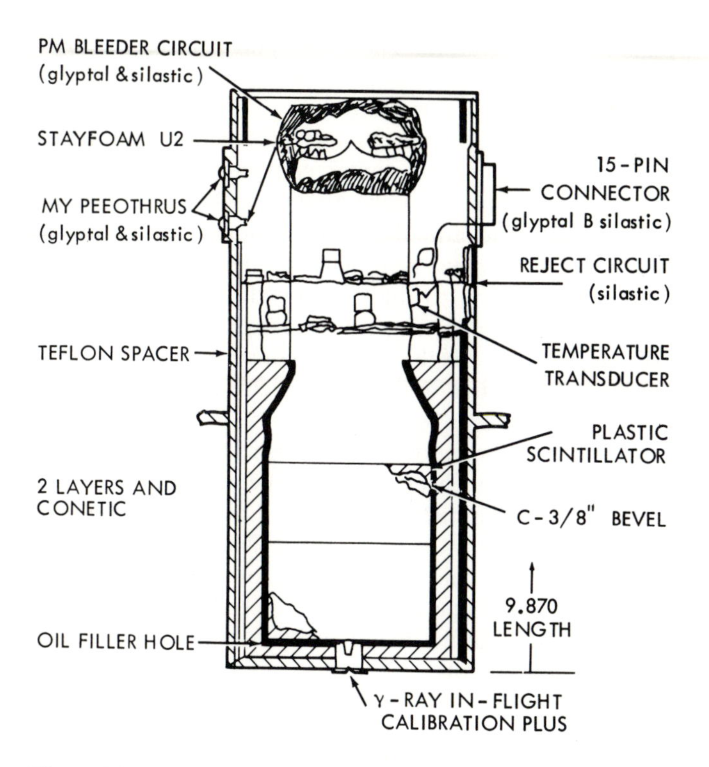

Figure 2.22 Cross sectional diagram of lunar spectrometer. Ranger mission. (Van Dilla, 1962)

plished in the following manner: A negative voltage at the anode of the photo-multiplier tube is fed into a network which has a 0.2 microsecond time constant. This network transmits the fast plastic scintillator pulse, but attenuates the slower cesium iodide pulse. The fast signal triggers a square wave pulse which inhibits the normally self gating pulse height analyzer. If a reject pulse is not generated, the signal derived from the eighth dynode of the photomultiplier (the gamma ray event) is analyzed in the 32 channel analyzer where the first channel accumulates all the overflow pulses of height greater than those analyzed in the following 31 channels. Details of the electronic design and the methods of packaging the equipment can be found in the references, Van Dilla (1962).

From all indications (Vinogradov et al., 1966), a very similar type of instrumentation was used on Lunas 10 and 11 except for the following two major differences: first, a smaller NaI(*Tl*) crystal (i.e. 30 × 40 mm) was used; and, second, no boom was employed to reduce the effects of the cosmic ray induced background. The development of rather high reliability, fast phoswhich circuitry made it possible to use NaI(*Tl*) crystals with their 0.25 microsecond rise times.

The analyzer which was part of the Ranger package (including the power supplies, amplifier, analog to digital converter, memory, and programmer) occupied a volume of a cube about 6 inches on edge and weighed about 5 pounds. The amplifier had an input sensitivity of 2.5×10^{-4} coulombs and a nominal gain of 100, providing a linear 5 V pulse height output. The analog to digital converter had an integral linearity of better than 3% over a temperature range of -10 to 55 °C and employed the usual pulse height to time conversion to generate digital storage address. Dead time was $10+2n$ microsecond where n is the channel in which the pulse was recorded.

Since the Ranger flight to the moon required 66 h, it was necessary to provide in-flight calibration to check on instrumental drift. A small calibration source was therefore included on-board to provide in-flight calibration. The sources were ^{57}Co and ^{203}Hg. These sources emit gamma rays of 73 kev, 112 kev, and 279 kev. The low energies emitted did not interfere with any of the true signals of interest. It was important for this type of experimental that no low intensity lines between 0.5 and 3 MeV be present as a background.

The pulse amplifier between the detector and analyzer had a two position gain switch providing two energy ranges for analysis: 0.02 to 0.6 MeV, and 0.1 to 3.0 MeV. From all available information about the Lunas 10 and 11, a similar analyzer was used. Calibration, however, was accomplished by monitoring environmental and operational parameters such as temperature, high voltage, and photomultiplier tube performance. Comparisons were made between the flight data and ground based experiments.

Ranger 3 Results

The Ranger 3 spacecraft did not pass close enough to the moon to perform its major function of measuring gamma radiation from the moon. However, more than 40 h of cis-lunar measurements were made and transmitted.

The first twelve hours of operation were performed with the detector close to the main structure of the vehicle; then, by means of a telescoping boom, the detector was moved to a position six feet away from the spacecraft. This operational procedure was used to determine radiation induced in the spacecraft. Alternate spectra were taken covering two energy ranges: 0.1 to 3 MeV (low gain), and 0.02 to 0.6 MeV (high gain).

There were known sources aboard the spacecraft, i.e., ^{60}Co and radium, in the accelerometers and radiometers. Correction was made for the effect of these sources, previously determined by ground-based measurements, and the data presented here reflect these corrections.

The gamma ray intensity in the stowed position was found to be less than a factor of two higher than in the extended position. Because the reduction in solid angle was about a factor of 20, Arnold et al. (1962), concluded that the outboard spectrum was essentially free of secondary radiation from the spacecraft. A low gain spectrum showing an average of 87 separate spectra is shown in Figure 2.23.

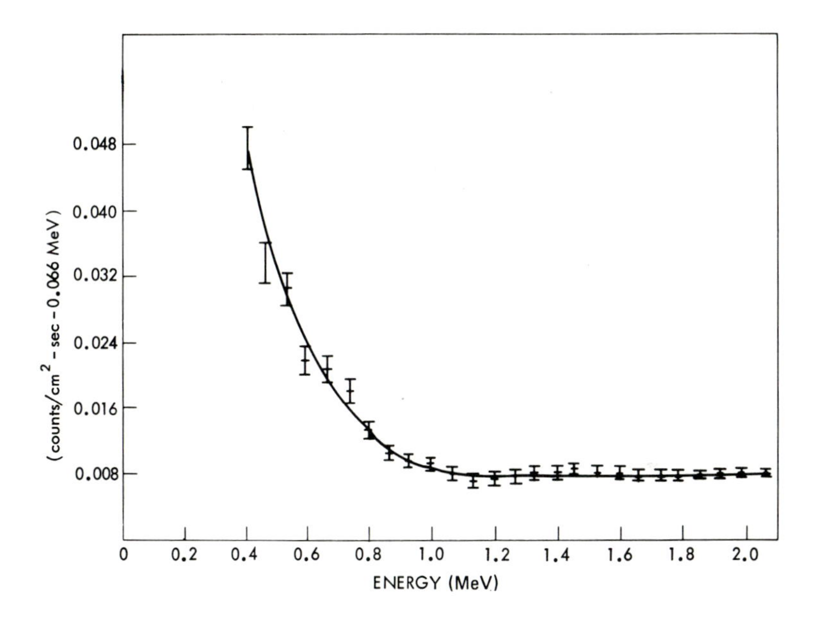

Figure 2.23 Averaged low-gain spectrum obtained in cislunar space by the γ ray spectrometer on Ranger 3 with the detector at a distance of 6 feet from the spacecraft. The bars represent $\pm 2\sigma$. (Arnold, 1962)

The total count rate between 0.5 MeV and 2.1 MeV was about 0.27 ± 0.01 counts/cm^2 sec, and about 0.67 ± 0.02 counts/cm^2 sec above 2.1 MeV (Arnold et al., 1962). Because of an unexpected temperature rise before boom extension, the gain changed in the extended portion of the experiment so that the actual energy region covered ranged from 0.1 to 2.1 MeV.

Luna 10 Results

On April 13, 1966, Luna 10 was put into an elliptic orbit about the moon. In this first period the minimum distance of the satellite from the moon was about 350km, the maximum distance about 1015km. The angle of inclination of the orbit, with respect to the lunar equator, was about 72 degrees. Figure 2.24 shows typical pulse-height spectra obtained with the gamma ray spectrometer on Luna 10. Curve **a** was the primary gamma ray pulse height spectrum near the moon; curve **b** was the background

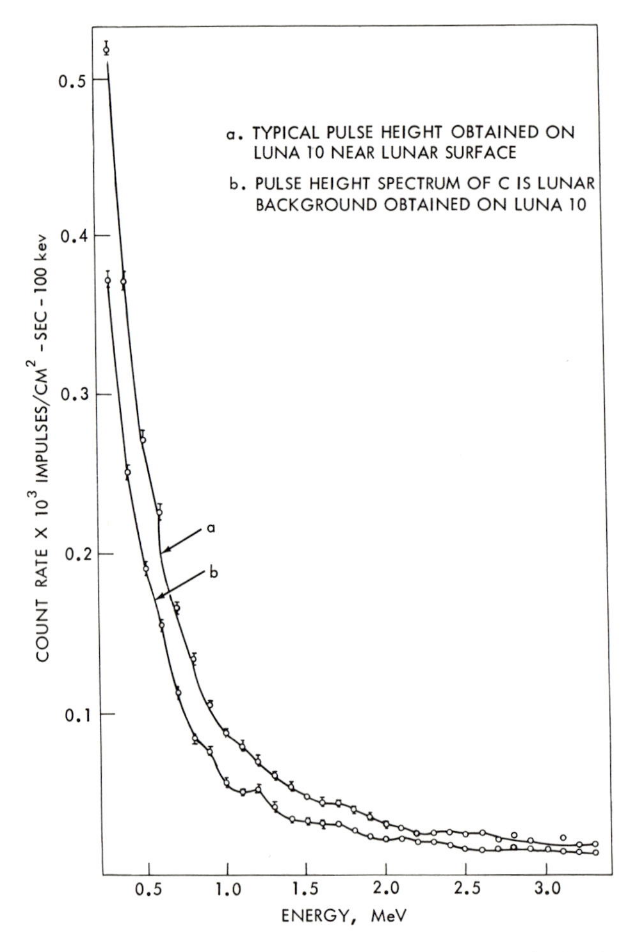

Figure 2.24 Gamma ray spectra, Luna 10. (a) Typical pulse height obtained on Luna 10 near lunar surface, (b) pulse height spectrum of cislunar background obtained on Luna 10

pulse height spectrum obtained during cis-lunar flight. The background spectrum was modified in absolute intensity by taking into account the cosmic radiation by the moon.

The problem was to determine two factors from the measured pulse height spectra: first the total gamma ray flux from the lunar surface (i.e., both induced and natural); and, second, to determine the contribution of the natural activity due to the ^{40}K, uranium, and thorium. The total gamma ray flux from the lunar surface was determined most easily. This contribution was just the difference between curve **a** and **b** in Figure 2.24. Figure 2.25 curve (**c**) is the pulse height spectrum of the

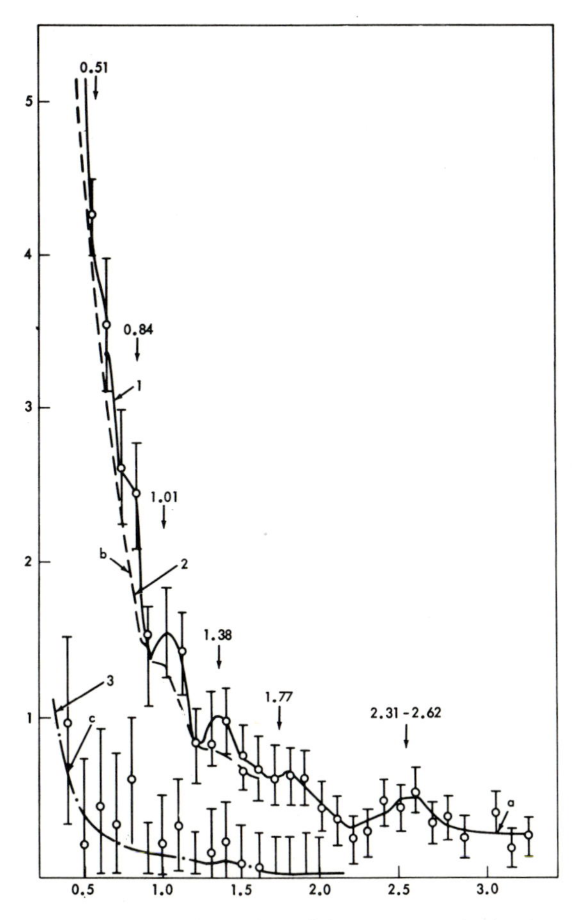

Figure 2.25 Typical pulse height spectrum obtained on Luna 10. (a) Primary pulse height spectrum, (b) background spectrum, (c) spectrum of natural lunar activity. (Vinogradov et al., 1967)

difference. The following assumptions were made in order to estimate the gamma ray flux caused by the natural radioactive elements:

1. The shapes of the induced gamma ray spectrum from various sources are similar (e.g. spacecraft, lunar surface).

2. The gamma ray flux above 2.5 MeV is due primarily to cosmic ray induced activity.

The cis-lunar background pulse height spectrum was used as the shape of the induced spectrum, and normalized so that its intensity at 2.5 MeV and greater was set equal to the intensity observed from the moon (Figure 2.25, curve **b**). The difference, then, between curves **a** and **b** was the contribution of the natural radioactivity (Figure 2.25, curve **c**).

On the basis of the above techniques and assumptions, the following conclusions were drawn by Vinogradov et al. (1967).

1. Not more than 10% of the total intensity of gamma radiation of lunar rocks can be attributed to natural radioactivity, whereas 90% of the observed gamma radiation is induced by incident cosmic radiation.

2. The intensity of gamma radiation above the lunar maria was observed to be approximately 1.15 to 1.2 times higher than that above the "mainlands".

3. The calculated intensity of the natural activity indicates that lunar surface materials correspond to terrestrial basalts. This can be taken as indication that the lunar mare material represents lava flows of basic composition.

References

Arnold, J. R.: The gamma spectrum of the moon's surface. Proceedings Lunar and Planetary Exploration Colloquium, 1958.

— Metzger, A. E., Anderson, E. C., Van Dilla, M. A.: Gamma rays in space, Ranger 3. J. Geophys. Res. **67/12,** 4878 (1962).

Franzgrote, E. J., Patterson, J. H., Turkevich, A. L.: Chemical analysis of the moon at Surveyor VII landing site, preliminary results. Jet Propulsion Laboratory Technical Report **32/1264,** 241 (1968).

Konstantinov, V. P., Bredov, M. M., Viktorov, S. V.: On the analysis of chemical composition of the moon's surface by direct methods. Doklady A. N. SSSR, Fizika, **181/4,** 827, (1968), „Nauka".

Morozov, A. A., Smorodinov, M. I., Shvarev, V. V., Cherkason, I. I.: Density measurements of the moon's surface layer by automatic station "Luna 13". Doklady A. N. SSSR, Tekhnicheskaya Fizika, **179/5,** 1087, (1968), „Nauka".

Negus de Wys, J.: Lunar surface electromagnetic properties. Magnet experiment, Jet Propulsion Laboratory Technical Report **32/1246,** 151 (1967).

— Electromagnetic properties. Magnet test, Jet Propulsion Laboratory Technical Report **32/1262,** 155 (1968).

Patterson, J. H., Turkevich, A. L., Franzgrote, E.: Chemical analysis of surfaces using alpha particles. J. Geophys. Res. **70/6,** 1311 (1965).

Rutherford, E., Chadwick, J., Ellis, C. D.: Radiation from Radioactive Substances. Cambridge University Press, 1930.

Scott, R. F., Roberson, F. I.: Soil mechanics surface sampler. Jet Propulsion Laboratory Technical Report No. **32/1264,** 135 (1968).

Turkevich, A.: Chemical analyses of surfaces by use of large angle scattering of heavy charged particles. Science **134,** 672 (1961).

Turkevich, A. I., Franzgrote, E. J., Patterson, J. H.: Chemical analysis on the moon at Surveyor VI landing site, preliminary results. Jet Propulsion Laboratory Technical Report **32/1262**, 127 (1968).

Van Dilla, M. A., Anderson, A. E. C., Metzger, A. E., Schuch, R. L.: Lunar composition by scintillation spectroscopy. I.R.E. Trans. Nuclear Science NS–**9,** 105 (1962).

Vinogradov, A. P., Surkov, I. A., Chernov, G. M., Kirnozov, F. F.: Measurements of gamma radiation of the moon's surface by the cosmic station Luna 10. Geochemistry no. 8, p. 891, V.I. Vernadsky Institute of Geochemistry and Analytical Chemistry, Moscow, USSR, 1966.

Chapter 3: Instruments and Techniques Under Development

We have discussed in detail, analytical, remote, automated, techniques that have performed successfully in flight missions. In this chapter, we will describe a variety of instruments in various stages of development. The principle purpose of these instruments, again, is to supply data for compositional analysis. These devices are under development for various types of missions (manned, unmanned and for surface and orbital explorations). Some have been proposed for lunar exploration, others for planetary studies; some techniques as we will see have obvious applications to both types of missions. The experimental devices involve different degrees of complexity, and consequently some are more optimal than others. The instruments to be described will be considered in groups as follows: 1. those using or measuring electromagnetic radiation (e.g. X-rays, gamma rays, infra-red, etc.), and 2. mass spectroscopy, gas chromatography, and a combination of both.

X-Ray Spectroscopy

The first method, X-ray spectroscopy, is potentially one of the most useful techniques for obtaining compositional information about lunar and planetary surfaces. Instrumentation can be relatively simple, the information obtained highly specific, and the data can be interpretated with relative ease. The actual development of X-ray analytical devices began with the initiation of the Surveyor program. The instrumentation continues to evolve, pointing towards flight opportunities involving extended lunar exploration and Martian soft landers.

In principle, an X-ray spectrograph consists of a means for exciting a characteristic X-ray line spectrum, a method for sorting and identifying the characteristic lines, and, finally, some method of detection and measurement.

Excitation of characteristic X-ray lines results from those interactions which produce inner shell vacancies in the atoms making up the samples under examination. The filling of these vacancies by outer shell electrons produces the emission of a characteristic X-ray line spectrum. Thus, transitions from outer atomic shell ending at the K level yield the K spectrum, while transitions to the L, M, etc. level produce corresponding

L, *M*, etc. spectra. In practice, characteristic X-rays are generated from a target by irradiating it with an electron beam, X-rays, or high energy charged particles such as protons or alpha particles. A necessary requirement is that the energy of the bombarding source be in excess of the binding energies of the electrons in the inner shells. As we will see, exciting radiation is available produced either by special apparatus or from radionuclides.

Once a characteristic X-ray line spectrum is produced, specific lines for measurement must be separated. This is done in order to identify the element from which a given line is emitted, and to measure the elemental concentration. There are alternative means for performing energy separation. The methods involve either X-ray optical means or electronic discrimination (pulse height analysis).

In the X-ray optical method, the practicing spectrographer uses a system involving a Bragg crystal monochromator and techniques based on Bragg's law of X-ray diffraction. The diffraction of X-rays by crystals is one of the most significant properties to the spectrographer. Crystals can be used in the manner of a three-dimensional diffraction grating, according to the Bragg's law expression:

$$n\lambda = 2d \sin\theta,$$

where n is the order of diffraction, λ is the diffracted wavelength, d is the distance between the lattice planes in the crystal, and θ is the incident and reflected angle of the X-ray beam to the lattice planes. It follows from the above that, for any given crystal with a given interplanar spacing, a particular wavelength or multiple will diffract at one and only one angle.

The method of electronic or pulse height analysis depends entirely on the use of energy sensitive radiation detectors such as proportional, scintillation, or solid state detectors. These detectors yield pulse amplitude distributions which are proportional to the energies of the incident radiation. An analysis of the pulse amplitudes by pulse height analysis methods gives information about incident energies. There are a variety of such detectors with distinctive characteristics, and their choice is dictated by specific applications. For space applications the requirements include such factors as stability, ruggedness, resistance to radiation damage, etc.

A simple comparison of the two methods of X-ray spectroscopy: the crystal spectrometer is generally more effective in providing energy resolution; pulse height analysis, on the other hand, lends itself to simpler instrumentation and substantially higher efficiencies.

The principles just described have been used to design several X-ray spectrographs to be used in flight experiments. These are discussed below.

JPL Lunar X-Ray Spectrograph (see Metzger, 1964)

The JPL lunar spectrograph was designed by Philips Electronics Instruments, Inc. to perform X-ray emission compositional analysis on the lunar surface as part of the Surveyor program. The device was built and delivered to JPL as a prototype breadboard for evaluation (see Figure 3.1). X-ray excitation is provided by means of a triode gun consisting of a hot tungsten filament, a control grid for beam focusing, and an accelerating anode. The filament is maintained at high negative potential, the control grid at 50 to 100 V positive with respect to the filament, and

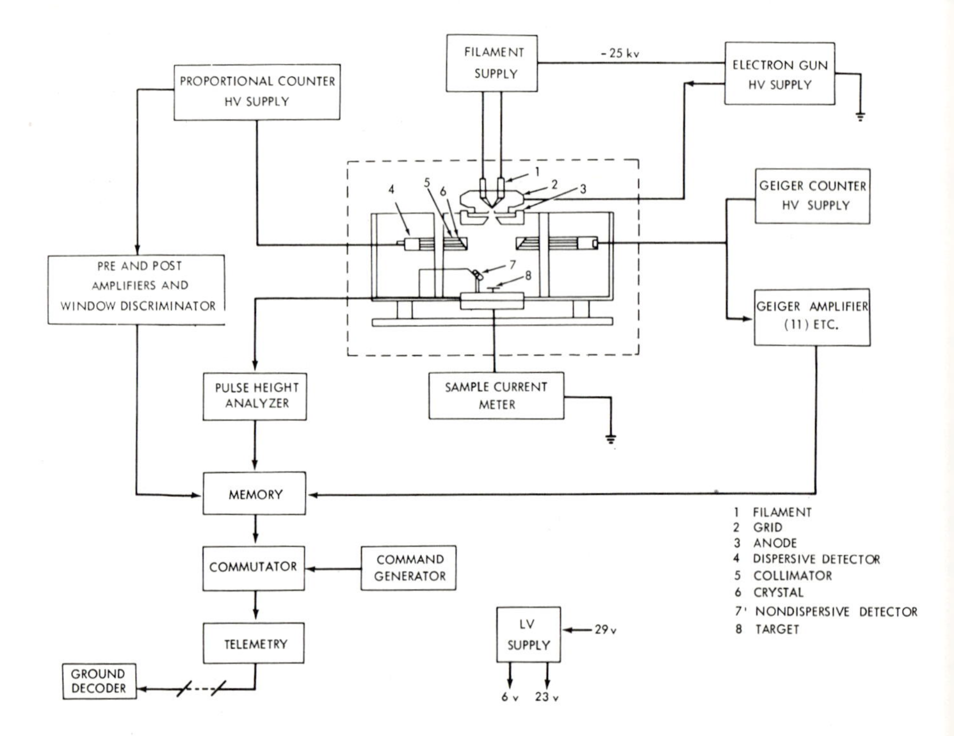

Figure 3.1 Schematic of the lunar X-ray spectrograph

the anode at ground potential. The electron gun is used to produce a beam of electrons incident on the sample through the hollow anode. The emitted X-rays are then viewed by 13 analysis channels, each tuned to a wavelength characteristic of an element of geochemical interest. The selected elements ranging from sodium to nickel are shown in Table 3.1.

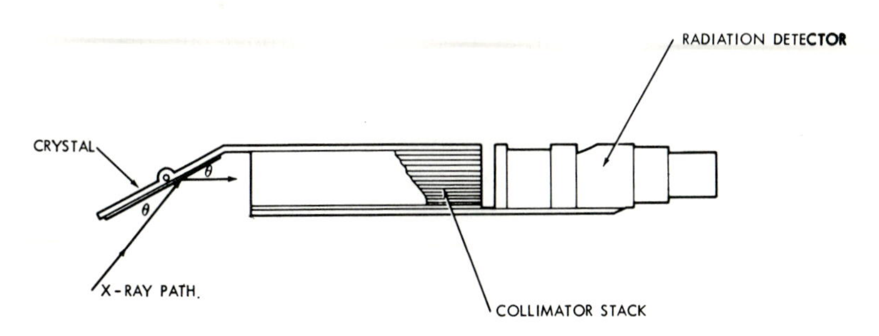

Figure 3.2 Optical discrimination of X-rays. (After Metzger,1964)

The details of the individual analyzers are shown in Figure 3.2. These are essentially small Bragg type spectrometers. The choice of the analyzing crystal and its angular setting with respect to the sample and detector is such as to respond to a single wavelength (corresponding to the elements listed in Table 3.1). The spectrometers consist of an analyzing crystal, a collimator stack of parallel plates (to ensure that only those X-rays, satisfying the Bragg reflection condition reach the detector), and a radiation detector such as an end-window proportional or a geiger counter.

A variety of analyzing crystals are used, depending on the selected wavelengths. These are NaCl, EDDT (ethylene diamine di-tartrate), and potassium acid phthallate. Geiger counters are selected for elements of atomic number 16 and above; end-window proportional counters are used for the lighter elements. The proportional counter outputs are fed to lower and upper level discriminator circuits in order to discriminate against higher order reflections, scattered radiation, and fluorescence from the crystals. One pulse height analysis channel is included among the spectrometers to view the sample directly. The detector, again, is a

Table 3.1. *Lunar X-ray spectrograph dispersive-channel capability*

Atomic number, Z	Element	Atomic number, Z	Element
11	Sodium	22	Titanium
12	Magnesium	23	Vanadium*
13	Aluminum	24	Chromium
14	Silicon	25	Manganese
16	Sulfur	26	Iron
19	Potassium	28	Nickel
20	Calcium		

* To be replaced by a chlorine channel.

proportional counter. Its output is processed through an analog-to-digital converter for pulse height analysis.

The purpose of this analyzer is to provide a backup to the crystal analyzer channels for identification of the major elements. It is included, also, to protect against failure of one or more channels, and to provide information on elements for which no prior programming has been done.

Specimen Requirements

Sampling is one of the major problems of this type of instrumentation. Samples must be collected, prepared by grinding and sieving, compacted, and then carefully presented to the electron beam for excitation. While this type of sample preparation is routine under laboratory conditions, it can obviously present very serious problems for a fully automated exploration mission. Metzger, in fact, reported problems of sample segregation where sample grinding and mixing was performed by machine. He, thus, found it necessary to shake the materials by hand. In addition to the problem of sample preparation, other problems were also identified and investigated. These included questions of sample conductivity, sample heating, heterogeneity, and the usual problems of X-ray analysis such as X-ray absorption and secondary excitation effects.

Sample Conductivity

Metzger has stated that, "sudden transients in electron beam flux can produce damage at the surface on many powdered samples, ranging from minor pitting to volcanic-like bursts of material". The effect of the electron beam on the target was observed to depend on the rate of rise of both

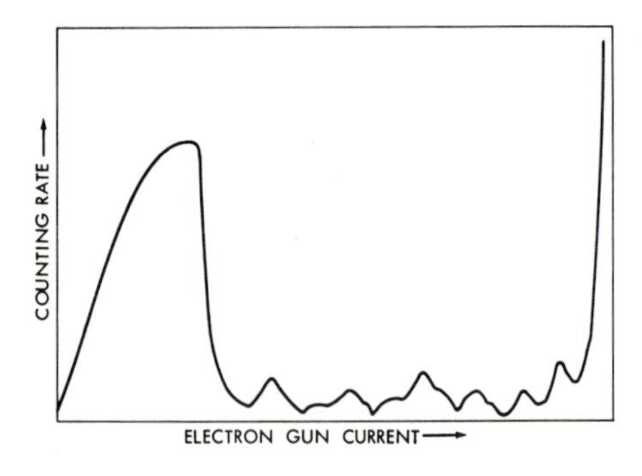

Figure 3.3 Response of a nonconducting powder to increasing electron gun current. (After Metzger, 1964)

the electron gun voltage and current. It was found necessary to slowly bring the gun to operating conditions in order to minimize surface damage. The effect of gun current on the observed X-ray emission is shown in Figure 3.3. There is an initial rise of count rate with gun current which then falls off with surface chargeup. The count rate remains low until a second sharp rise, attributed by Metzger to a sudden decrease in resistivity due to increased heating. This has been confirmed by resistance measurements. Various approaches have been taken to minimize these charging effects such as the use of a 40 mesh screen across the top of the graphite sample container cup, a 25 mil copper pin in the center of the cup, and, finally and most successful, the procedure of uniformly distributing powered graphite through the sample. The method ultimately chosen was to distribute about 15% of added graphite through the sample. This procedure adds additional difficulties to the problem of sampling under automated conditions, and would be complicated even with astronaut involvement.

Sample Heating

Measurements performed under conditions of maximum incident electron flux of 25 kev and 50 ma, delivered to a spot $\frac{3}{16}$ in. in diameter, show a temperature of about 230 °C. This is considered to be generally inconsequential, although the loss of volatile compents such as SO_2 and Cl_2 can be expected.

Compactness and Surface Smoothness

Metzger has observed that even with a conductive filler such as carbon it is useful to compact the specimen in order to maintain surface integrity. A value of 5 pounds per square inch is cited as being sufficient. An immediate benefit of increased surface smoothness is to minimize absorption effects, particularly for the lighter elements.

Rock Analysis

The performance of JPL lunar spectograph has been evaluated under laboratory conditions on four rock specimens: a sulfide, a syenite, a gabbro, and a granite. The results taken from eight of the crystal channels are shown in Figure 3.4. Intensities vs chemical concentration are plotted for Fe, Mn, Ti, Ca, K, Si, Al, and Mg. Included are original and corrected

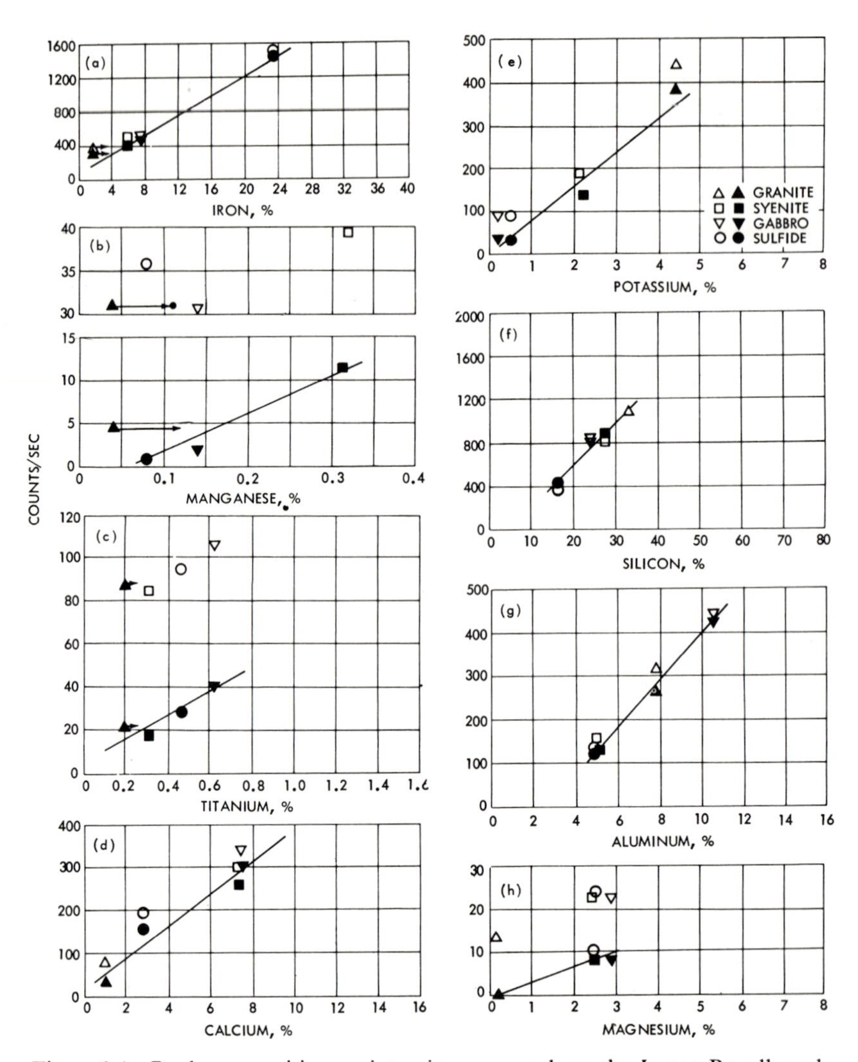

Figure 3.4 Rock composition vs intensity measured on the Lunar Breadboard X-ray Spectrograph. Unshaded symbols represent observed counting rates (adjusted for counter dead time and added carbon); shaded symbols represent observed (adjusted) counting rates corrected for background and enhancement, as required. (After Metzger, 1964)

count rates. These corrected count rates were obtained by subtracting a background obtained from looking at a silica sample with the various channels. No corrections were made for silicon and manganese. A weighted continuum correction based on the relative concentrations of iron and silicon in the rock was used.

Metzger concludes that the above experiments demonstrate that the instrument is capable of quantitative analysis for elements present in major and minor abundances. Sensitivities are 0.1% or better for all but the lightest elements of geochemical interest.

Goddard Space Flight Center-Panametrics Spectrometer

We have described different methods for performing X-ray spectroscopy, and discussed the JPL instrument built around electron beam excitation and crystal optics. We will now consider an alternative method based on the use of an alpha-emitter radionuclide for excitation and pulse height analytical methods. This device has been developed as part of a cooperative program between NASA's Goddard Space Flight Center and the Panametrics Corp., and is designed to meet such requirements as small size, reliability, stability, and specificity (Adler and Trombka, 1968). The major purpose for this development is twofold: 1. to provide a means to help the astronauts select samples for return, and 2. to obtain geochemical information for mapping and preliminary surface analysis for manned and unmanned lunar missions.

Principles

X-ray spectrometers employing radionuclide excitation sources, energy sensitive detectors, and pulse height spectral analysis have been proposed and, in some instances, developed by a number of investigators: Cameron and Rhodes (1961), Robert and Martinelli (1964) Friedman (1964), Kartunnen (1964), and Sellers and Ziegler (1964). Most devices have used gamma, beta, or bremsstrahlung sources, but only recently has a systematic study of alpha excitation been performed (Sellers and Ziegler, 1964; Robert, 1964; and Imamura et al., 1965).

The various sources have advantages and disadvantages depending on the nature of the application. Beta sources excite characteristic spectra in the same manner as the electron beam excitation already described in the technique used by Metzger. Because of the higher energies of most useful beta sources, a continuum of bremsstrahlung radiation is also produced which acts as a substantial background on which the characteristic X-ray lines appear.

As a variant, beta isotopes such as ^{3}H and ^{147}Pm have been incorporated into a metal matrix such as titanium or zirconium. These are known as bremsstrahlung sources. ^{3}H has a maximum beta energy of about 19 kev and, in combination with zirconium, the output spectrum is a continuous one between approximately 3 and 12 kev. In combination with titanium, the spectral output is a nearly characteristic one peaking near 5 kev. ^{147}Pm emits beta particles with a maximum energy of 225 kev, and thus gives a continuum with a much broader distribution. Excitation by these sources is by X-ray fluorescence. Backgrounds are much lower than by beta excitation, although interference is still encountered due to X-ray scatter. These sources have proved to have only limited usefulness in light element analysis because of low ionization cross sections.

Attention has recently been focused on the use of alpha excitation, which, as we will see, is particularly effective and useful in the specific program of analyzing and identifying rock types and the rock-forming elements Na, Mg, Al, Si, K, Ca, Ti, and Fe.

Chadwick (1912) was the first to note that bombardment of a target with high energy alpha particles produced characteristic X-radiation. It was observed that energetic alpha particles like electrons and X-rays are capable of producing ionizations in the inner electron shells of atoms. Merzbacher and Lewis (1958) published a review of the theories for X-ray production by alpha particle bomdardment. It was shown that the ionization cross sections for heavy charged particles vary approximately as the inverse of the sixth power of the atomic number, and directly as the square of the alpha particle energy.

The implication for the analysis of light elements is: that the rapidly increasing ionization cross section with decreasing atomic number competes with the decreasing fluorescence yield and detector efficiency in such a manner as to produce a maximum in the characteristic X-ray yield for the light elements; the actual position of the maximum depends on the thickness of the detector window.

A summary of the useful properties of alpha particles for exciting the low atomic number elements has been given by Sellers et al. as follows:

1. The cross sections for ionization of the K shell electrons increase approximately as the inverse of the sixth power of the atomic number of the excited element, and directly as the square of the alpha particle energy. This rapidly increasing ionization cross section with decreasing atomic number compensates well for the decreasing fluorescence yields.

2. In comparison to electron excitation, alpha particles produce a negligible bremsstrahlung continuum. The continuum is down by a factor of $(m/M)^2$ where m and M respectively are the masses of the electron and alpha particle.

3. Although alpha particle emission is usually associated with gamma ray emission, it is possible to obtain sources where gamma ray radiation is both a small fraction of the total emission and the gamma ray wavelengths sufficiently removed from the generated X-rays to be easily distinguished from them.

4. Practically, it is possible to acquire sources of high specific activity that emit very energetic alpha particles having energies of 4 to 6 MeV such as ^{210}Po or ^{242}Cm. However, the phenomenon of aggregate recoil and the consequent health hazard make it necessary to use sealed sources. Commercial sealed sources of ^{210}Po with high activity, enclosed by ultra-thin foils of stainless steel or platinum, can now be obtained. Although these windows do degrade the alpha particle energies, the loss of energy is not a practical problem.

In paragraph 2 above, we stated that alpha bombardment produces virtually no bremsstrahlung. Nevertheless, there is some observable background that comes from X-rays generated in the materials of the source container and window. Additional sources of background may be due to processes of internally converted gamma rays in the radio-nuclide, and X-rays emitted by contaminating radioactive species. It has been further stated by Sellers that, because of the decrease in cross section for X-ray production with increasing atomic number, the yield for elements like Fe would be expected to be low from pure alpha excitation. Compensation can be achieved by the use of a combined alpha and X-ray source, relying on excitation of the heavier elements by the emitted X-rays. This can be accomplished by the choice of a proper substrate on which the alpha emitting nuclide is deposited, as well as the choice of a suitable source window. The alpha particles produce X-rays in the substrate or window which, in turn, produces fluorescent X-rays from the heavier elements. The major consideration is to obtain exciting X-rays with energies greater than the absorption edge energies of the elements being determined.

Table 3.2 gives some experimentally determined X-ray yields expressed as X-ray quanta/alpha-steradian. These values have been determined by Sellers for a number of pure elements using a polonium source having the following characteristics: 0.3 millicuries activity deposited on a platinum backing, and a 0.17 mil window. The figures shown in the table indicate the theoretically expected, rapidly increasing yield with decreasing atomic number.

Table 3.2. *Measured alpha excited* X-*ray yields. Energy of the alpha particles approximately* 2 MeV. *Source strength approximately* 0.3 mc.

Element	Z	Yield (X-rays/alpha-steradian)
Al	13	4×10^{-4}
Ti	22	1×10^{-5}
Cu	29	9×10^{-6}

(After Sellers, 1964)

Instrumentation

Figure 3.5 shows the arrangement of the various components of the spectrometer head. The apparatus consists of an alpha source for excitaction of the characteristic spectra, a thin-windowed sealed proportional counter, and a head unit for mounting the source and detector and through

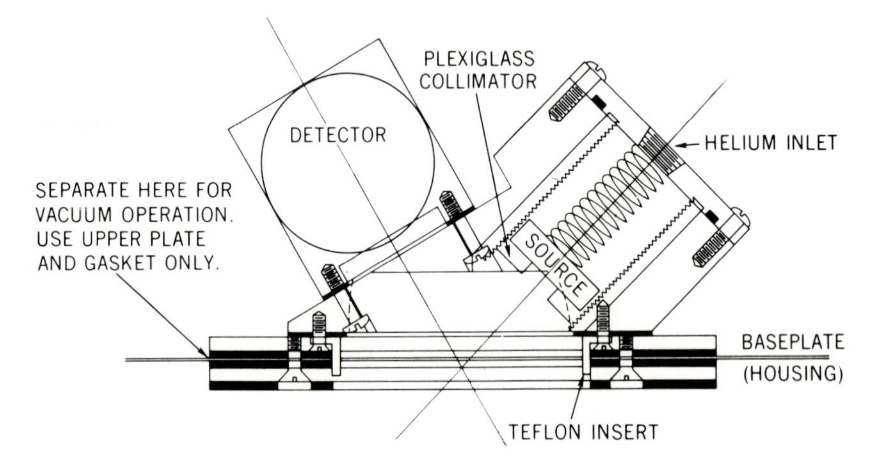

Figure 3.5 Optimum geometry lunar analyzer head assembly drawing

which the alpha particles are directed towards the sample which, in the lunar exploration mode, would be the surface. The total volume of this analyzer head is of the order of about 64 cubic inches, and the weight about 2 pounds. A description of the components follows:

Detectors

Since the concern is with the measurement of X-rays down to 1 kev, the detector window must be very thin and have a transmission at these low energies as well as a high efficiency and the best possible energy resolution. The best detector available at this time is the thin windowed proportional counter, either flow or sealed. An attempt has been made to work with windows of 0.001″ Be or 0.00025″ Mylar. Figure 3.6 shows a typical efficiency curve for a 2 cm detector using $Ar—CH^4$ and a 0.001″ Be window. This curve represents a product of window transmission and X-ray absorption. It can be seen that the counter efficiency is still about 10% at 1 kev. (The combination of low efficiency and low fluorescence for the low atomic number elements makes the high ionization cross

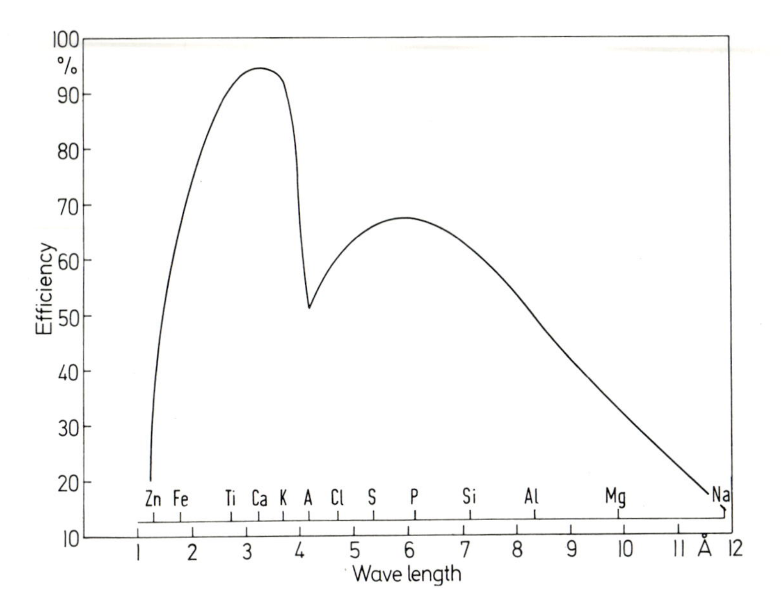

Figure 3.6 Detector efficiency: 1 cm proportional counter, Ar—CH_4, 0.001 in. beryllium window

sections of the alpha particles particularly significant.) The detector energy resolution is also a very important parameter in instrument design. One can now obtain detectors with about 18% resolution for the Mn K line from ^{55}Fe.

Electronics

The electronic system is of a conventional type consisting of a battery operated DC to DC converter high voltage supply, a high quality-low noise charge sensitive preamplifier, and a multichannel analyzer. Future plans will involve the use of a flight configured linear amplifier and a 2 MHz analog to digital converter.

Data Analysis

Figure 3.7 demonstrates the character of the observed pulse height spectrum given by two different rocks, dunite and granite. A number of peaks may be observed which are characteristic of certain energy regions. The envelope in region 1 is made up of contributions by the K α X-ray lines from Na, Mg, Al, and Si; region 2 by the K α lines from K and Ca;

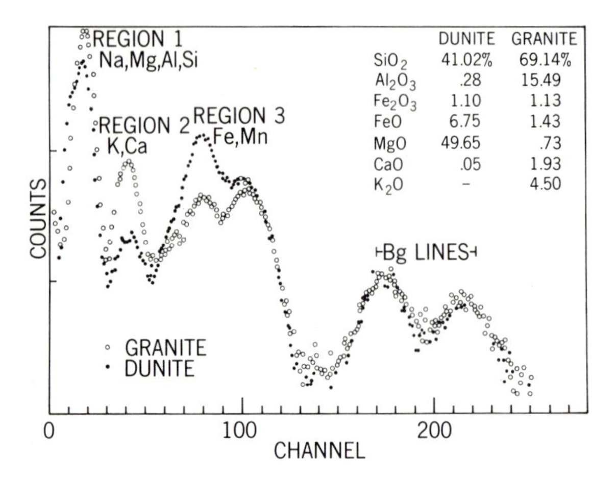

Figure 3.7 Observed pulse height spectra for a dunite and granite

and region 3 by Fe, Mn, and Ni. The other peaks are caused by background lines originating in the source (in this instance ^{242}Cm) and the source holder. Thus, the total pulse height spectrum is an envelope made up of the various components in varying proportions.

The method of data reduction must reduce the observed pulse height spectra to both qualitative and quantitative information. This can be done in two sequential steps: 1. the differential X-ray energy spectrum is determined from the pulse height spectrum; and 2. the relative chemical compositions are deduced from the energy spectrum.

The method for performing this type of data reduction is a linear-least square analysis which performs these functions simultaneously. This program, under development at the Goddard Space Flight Center, has been described by Trombka and Schmadebeck (1968). It will be treated in details in chapter 5. It allows for the determination of the presence of a spectral component, the relative concentration, the calculation of statistical error, gives an estimate of goodness of fit between the calculated and observed spectrum, and compensates for gain and zero drift.

Figure 3.8 shows an example of spectrum fitting using the above program to analyze a spectrum obtained from a sample of dunite. The open circles are the observed data, and the solid line represents the computer results. Once again regions 1, 2, and 3 define the energy regions of interest. The remainder is the computer fit to the background. This region is somewhat more difficult to fit because the background comes from scatter which is sensitive to the total composition, and is less well known.

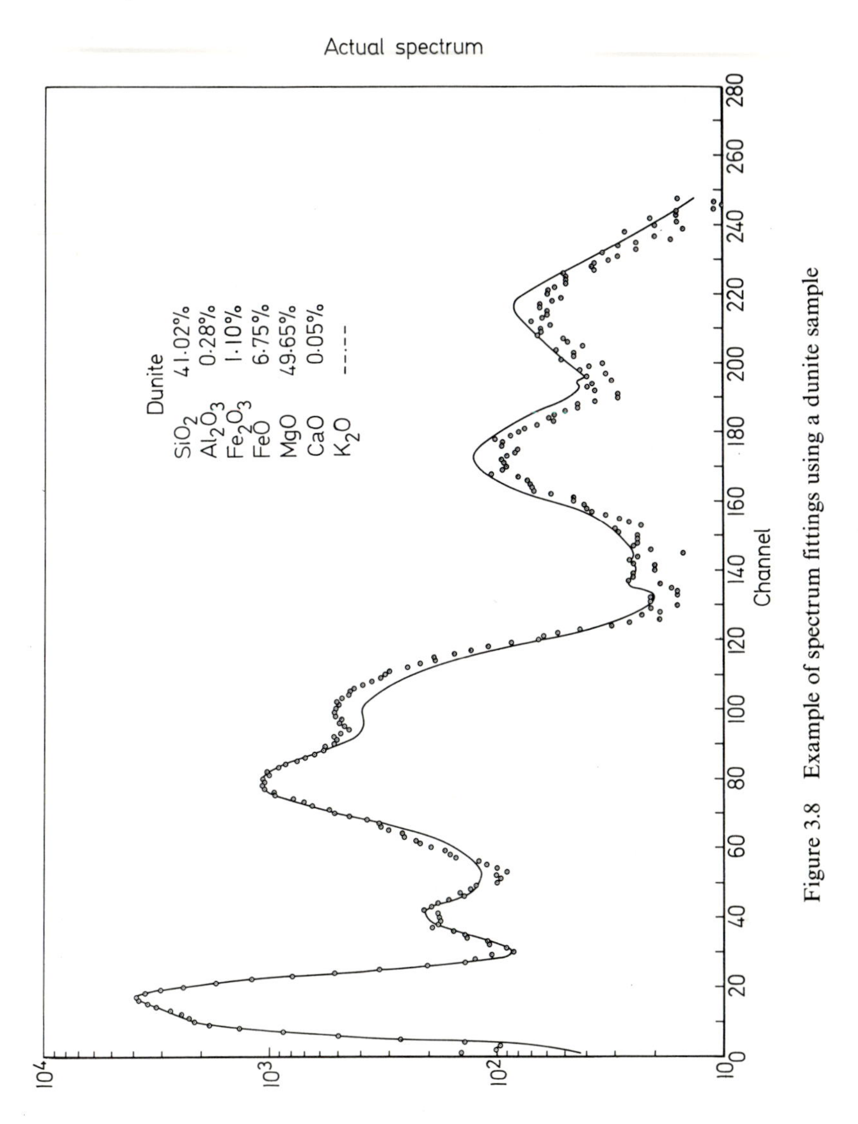

Figure 3.8 Example of spectrum fittings using a dunite sample

Some examples of the possible application of the overall technique to rock identification are shown in the following figures where the known chemical ratios are plotted against the experimentally determined intensity ratios. In all cases, normalization was against silicon. Figure 3.9 a shows Fe/Si curves for five different rocks ranging from acidic to ultrabasic. A similar curve for Ca is shown in Figure 3.9 b. This can be done

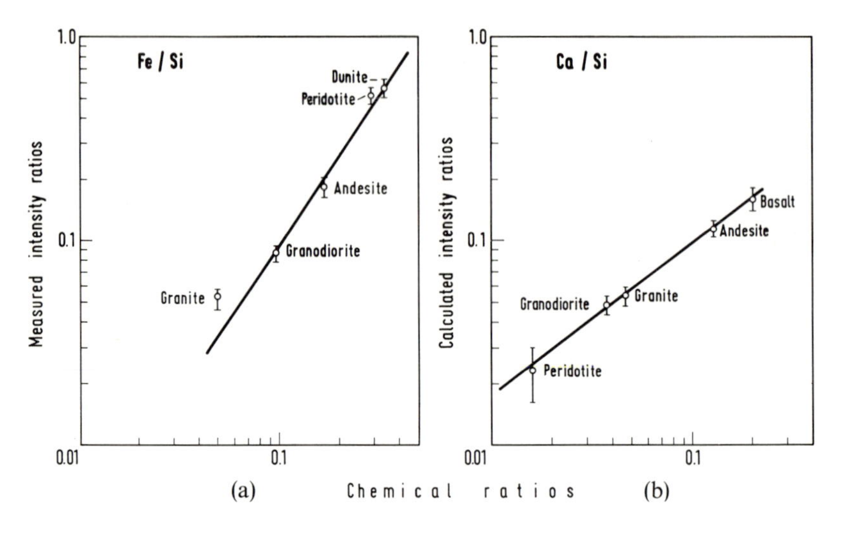

Figure 3.9 Fe/Si curves for five different rocks ranging from acidic to ultrabasic

for all the elements of interest. The slopes will vary and the order of rock types will be different for different elemental ratios; but if the range of variation of elemental ratios for a given rock type can be established, and a number of criteria for comparison can be set up, then a rock type could be identified with a substantial degree of certainty. An effort, in fact, is under way to use the computer for this type of operation.

Finally, Table 3.3 shows the results of a series of quantitative determinations on the Plainview Meteorite, using analyzed rocks and samples as standards. The results are considered very adequate for a possible lunar probe where the emphasis will be on sample selection and rapid survey analysis.

Table 3.3. *Results obtained for Plainview Meteorite using the least square analysis*

Element	Relative intensities from least square analysis	% Composition from Figure 17	Reported chemical analysis*
Fe	$0.36 \times 10^5 \pm 0.02 \times 10^5$	14.1 $\pm$2.8	17.0–24.0
Si	$0.31 \times 10^5 \pm 0.01 \times 10^5$	17.5 $\pm$0.7	17.25
Al	$0.25 \times 10^4 \pm 0.09 \times 10^4$	3.2 $\pm$1.5	1.08
Mg	$0.54 \times 10^4 \pm 0.05 \times 10^4$	18.0 $\pm$2.5	13.71
Ca	$0.24 \times 10^4 \pm 0.03 \times 10^4$	1.08 $\pm$0.21	1.19
K	$0.95 \times 10^3 \pm 0.30 \times 10^3$	0.1 $\pm$0.03	0.066

* Quarterly rept of the US Geological Survey, April-June 1965.

X-Ray Diffraction Techniques for Lunar and Planetary Studies

The development of X-ray diffraction methods for lunar and planetary exploration was undertaken with the objective of learning about the mineral composition of lunar and planetary surfaces. The goal was to provide a flightworthy X-ray diffraction unit for unmanned as well as manned missions. Included in this program were the development of a sampler system to provide samples to the diffractometer unit, and a study of data interpretation to learn how effectively X-ray diffraction could be used (either alone or with some determination of elemental composition) to identify the kinds and amounts of different mineral phases present. Involved in this program were a number of investigators; H. H. Hess of Princeton, R. C. Speed, formerly with the Jet Propulsion Laboratory (now at Northwestern University), and a project team consisting of J. A. Dunne, N. L. Nickle, and D. B. Nash (all of the JPL). Additional investigations were carried out under the guidance of A. E. Metzger (see Das Gupta et al., 1966).

Historically, work on the diffractometer was begun by W. Parrish of the Philips Laboratories under the cognizance of R. C. Speed. A number of instrument models were built, some actually in flight configuration. While the original models were constructed around a conventional Bragg-Brentano geometry with scaled down dimensions, later models using Seemann-Bohlin optics were also investigated.

Principles

The X-ray diffraction method depends on the phenomenon of coherent X-ray scattering by crystalline materials, and is one of the most widely used techniques for identifying chemical compounds. The basis for the method is the Bragg law of diffraction: $n\lambda = 2d\sin\theta$, where λ is the wavelength of the diffracted X-rays, d is the distance between the diffracting planes, and θ is the angle between the incident beam and the diffracting planes. This law dictates that diffraction occurs only from those lattice planes having the proper lattice spacing, d, at a given angle, θ, to satisfy the above expression. When dealing with powdered samples, one expects that a great many of the small, randomly oriented crystals will have the proper orientation to participate in diffracting an X-ray beam of a suitable wavelength at an appropriate incident angle.

An instrument for performing diffraction measurements requires: a source of essentially monochromatic X-radiation for irradiating the sample, a sample holder, a means for detecting the diffracted radiation, and a precise means for determining the diffraction angle. Two major types of X-ray optics have been investigated, Bragg-Brentano and Seeman-

Bohlin. Bragg-Brentano optics is the basis for the majority of commercial diffractometers (see Figure 3.10). Because the use of flat specimens in such optics introduces peak shape and peak position aberrations, Parrish introduced, as a refinement, a special, curved sample cup with a beryllium window to provide both sample containment and the necessary sample curvature to minimize the above effects. The geometry of the diffractometer was arranged so that the beryllium window faced downward. Additional experiments were also conducted on upright geometries (samples face up). As reported by Dunne and Nickle, this led to additional problems in sample preparation.

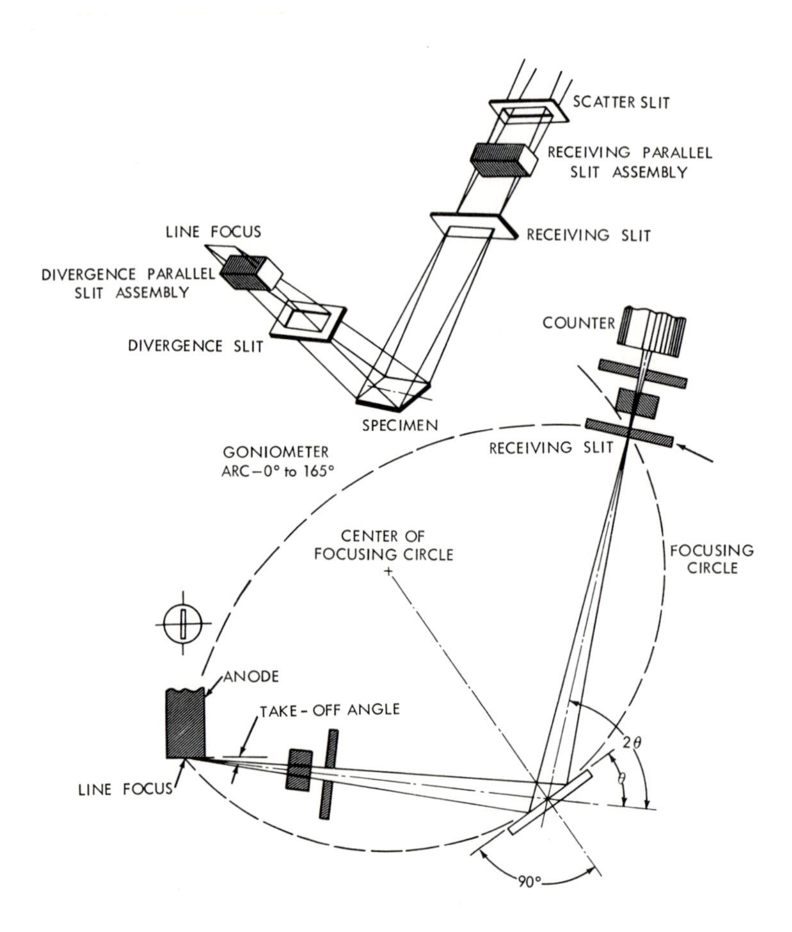

Figure 3.10 X-ray optical system of the Philips goniometer. (After Parrish, Hamacher, and Lowitzch, 1954)

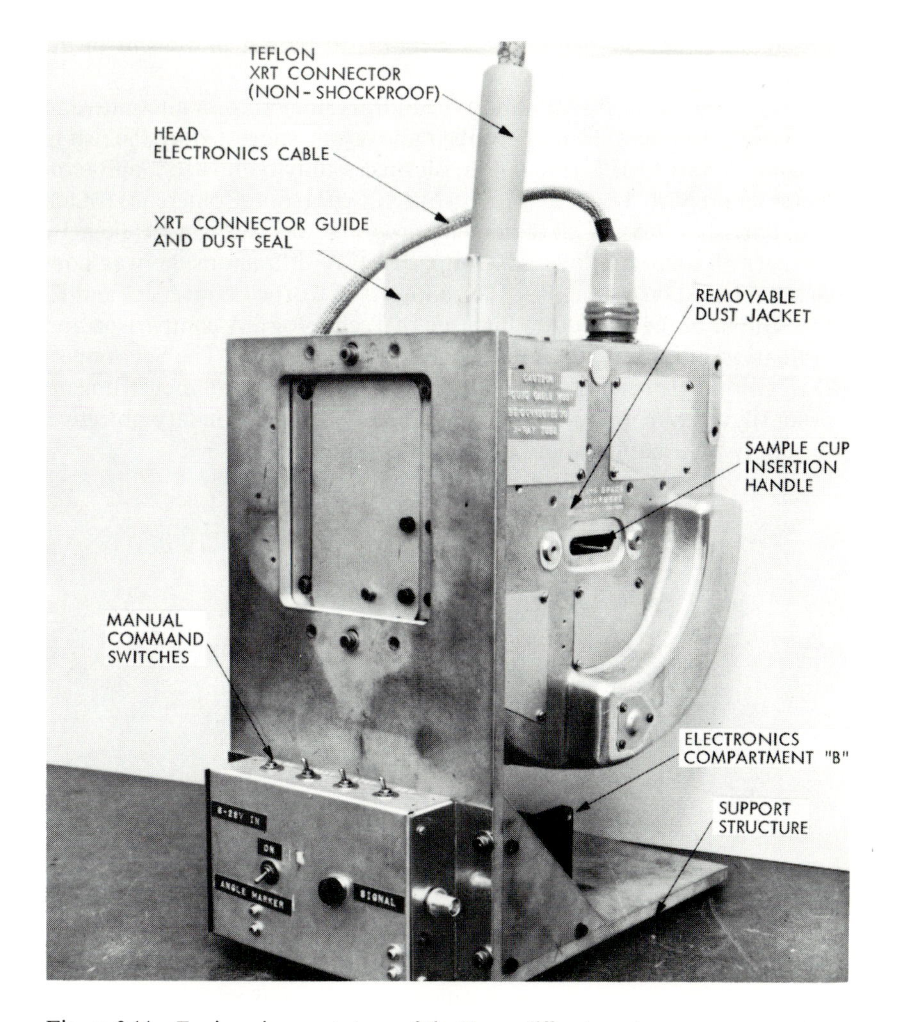

Figure 3.11 Engineering prototype of the X-ray diffractometer

Figure 3.11 shows an engineering prototype of a field operable diffractometer. Shown is the inserted X-ray tube, the command box, the electronics compartment, and the opening through which the sample is inserted. The X-ray tube designed for use with the diffractometer has a copper target and will operate at 25 kev and 1 ma. The tube has been designed to operate for an extended period without any special means for cooling. In a subsequent model, the X-ray tube was integrated into the high voltage supply thus eliminating high voltage connector and cabling problems. A flexible bellows was included to permit movement of the X-ray tube for alignment.

The use of a Seeman-Bohlin type diffractometer as an alternative to the Bragg-Brentano device described above was suggested by Parrish in the final report to JPL (1964). Parrish successfully built such a diffractometer and reported on it at the Pittsburgh Diffraction Conference (Mack and Parrish, 1965). A development program was also undertaken by Metzger (Radiation Physics Group, JPL). A diffractometer was constructed based on a design by H. Schnopper of Cornell University and K. Das Gupta of the California Institute of Technology. A comparison and evaluation of both types of instrumentation was made. The Schnopper-Das Gupta instrument was designed as a dual purpose instrument: primarily for mineralogical identification, and, as a secondary objective, to supply information about elemental composition.

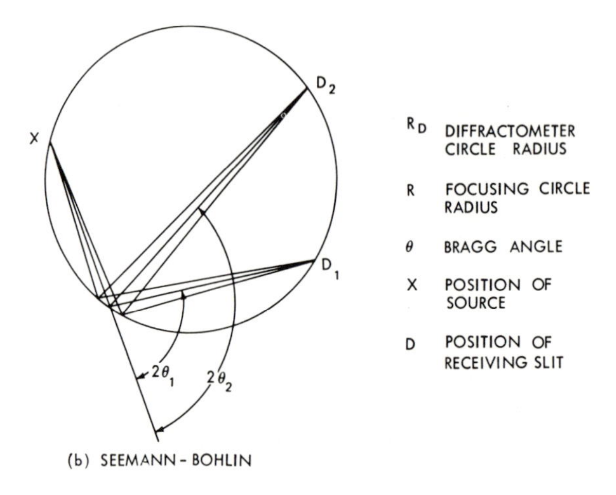

Figure 3.12 Diagrammatic representation of the Bragg-Brentano and Seeman-Bohlin X-ray powder diffractometer optical arrangements

Figure 3.12 is a schematic representation of the optics employed in the JPL Seeman-Bohlin instrument. The source slit, powdered sample, and detector slit all lie on the circumference of the focusing circle. The geometry is based on a fixed source-sample relationship so that a simple focusing design is possible. This involves placing a line source directly on the focusing circle as proposed by Das Gupta (1958). The line source on the focusing circle is imaged at various points on the focusing circle, corresponding to the various lattice planes as they diffract from a fixed, curved sample. In order to detect the diffracted radiation, the detector, by means of a simple linkage, is made to rotate about the center of the focal circle while being constrained to point a narrow receiving slit at the center of the sample. The instrument was also built to permit one or more proportional counters to directly view the sample for performing elemental analysis by pulse height analysis.

The following advantages have been claimed for the instrument:

1. High diffracted beam intensities per power input.
2. Mechanical simplicity: The number of moving assemblies were reduced from two to one.
3. Light weight.
4. The possible use of multiple detectors for rapid diffractometer scans.
5. Simultaneous diffraction and chemical analysis.

The one major disadvantage of the Seeman-Bohlin diffractometer is that, for mechanical reasons, the low angle diffraction is about 20 deg. 2 theta. Dunne has pointed out in this regard that while most of the important diagnostic lines of the major rock forming minerals occupy an angular region above 20 deg. (for Cu or Cr radiation), there are some significant exceptions (including sheet silicates and hydrous phases, both of which are of critical petrologic importance). Thus the limited angular range of the Seeman-Bohlin arrangement places the Bragg-Brentano optics in a much more favorable relative position despite its inherently lower efficiency.

Sampling and Sample Preparation

The acquisition and preparation of samples to be used in the lunar diffractometer is a major problem. It is obvious that the successful use of the lunar diffractometer hinges on the delivery to it of selected samples properly prepared. These samples have to be representative of the rocks and soils from which they come, and they must be suitable for diffraction measurements. For unmanned missions, one must build automated equipment for performing these functions. This subjects has been discussed in detail by D. B. Nash (1965).

In order to perform precise and reproducible X-ray diffraction measurements, it is necessary to consider such factors as particle size distribution, homogeneity, surface geometry, porosity, contamination, and orientation effects. It has been found, for example, that the optimum particle size when dealing with powders is between 1 and 10 microns; although, as Nash indicates, particles as large as 100 microns can yield useful results as long as there is a high proportion of fine particles. In the instrument using the inverted geometry (Be window facing downward), one can expect that the fine particles will sift to the bottom so that the X-ray beam would in effect see a higher fraction of fine powders.

Figure 3.13 shows the effect of particle size for quartz and basalt. The major effect traceable to orientation effects is a change in peak intensities. This phenomenon can seriously complicate the identification of phases. Additional particle size effects can result from the overgrinding of samples. It is well known that particles below 0.1 micron will cause diffraction peak broadening.

The above is to be considered a cursory review of the sampling problem. The reader is referred to the Nash article for a more complete discussion. It can be stated, however, that the use of the diffractometer

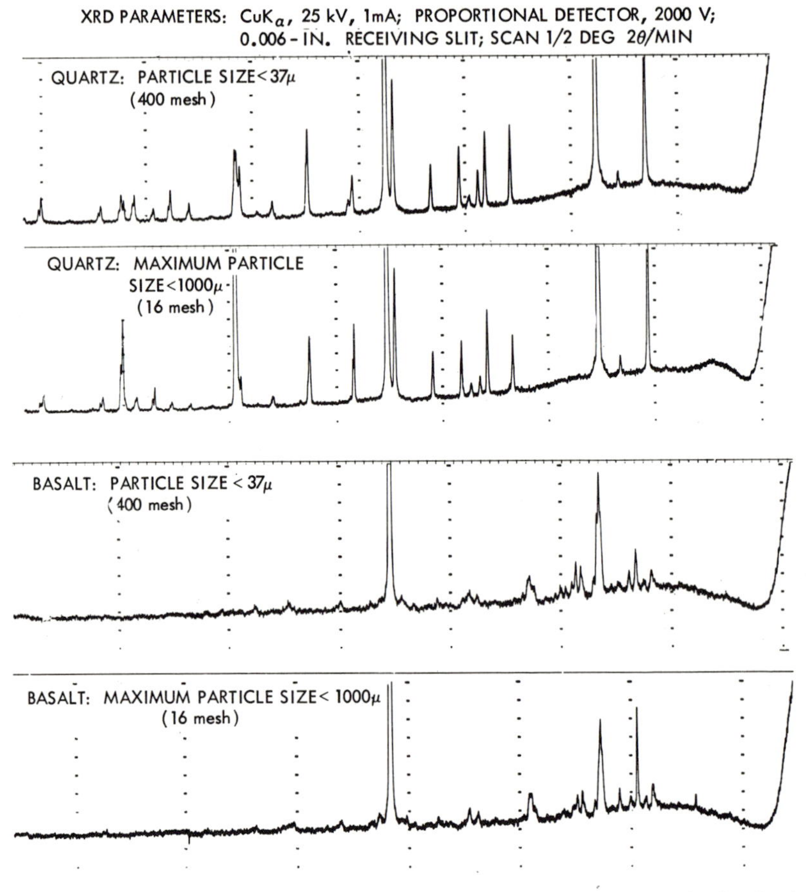

Figure 3.13 Effects of maximum particle size on diffraction patterns of pulverized-rock specimens

for lunar and planetary missions puts very difficult and serious requirements on sample collection and preparation. In point of fact, this aspect of the problem is one of the major limitations; not only must the sample be prepared, it must also be delivered to the instrument for measurement.

Data Handling and Interpretation

The subject of data acquisition and interpretation has been treated by Speed et al. (1965), and by Nash (1965). Both undertook a systematic study of the question of how effectively one can determine the nature of rocks, and the effects of instrumental and specimen variables. These studies were very comprehensive, involving over 500 rock samples, and are described in detail in the JPL *Technical Memorandum* 33-218. The conclusions have also been summarized in the same report by G. Otalora. The general conclusions drawn are that the X-ray diffractometer is uniquely suited for the determination of the chemical nature of crystalline rocks. Interpretation of the data follows either from visual inspection of the diffractograph or from a computer program for handling the data. Further, it is proposed that by examining and measuring selected diffraction lines, it will be possible to determine the relative abundance of mineral phases in the rocks.

Orbital X-Ray Fluorescence Experiment (Adler et al., 1969)

The following has been proposed as one of a group of orbital experiments for later Apollo missions to perform a compositional survey of the lunar surface. Its purpose is to measure the secondary X-rays produced in the lunar surface by solar X-rays. Solar X-rays interact with lunar surface materials to produce characteristic X-rays of the major elements making up the surface composition. The flux of these secondary X-rays is a function of the intensity of the exciting flux as well as its spectral composition. The intensity of the exciting flux, in turn, is related to the solar activity, and varies considerably during quiet, active, and flare periods.

The objective of the X-ray experiment is to measure the K spectral lines from the Na, Mg, Al, Si, K, Ca, and Fe from the lunar surface. From these measurements, one would hope to obtain the following information:

1. Data about the nature of a large portion of the lunar surface.
2. A measure of the chemical homogeneity of the lunar surface as the spacecraft sweeps around the moon.
3. By comparison with gamma ray results, some idea of the extent of gardening and whether the composition of the surface is like that of the immediate subsurface.

Experimental Approach

The experimental approach consists of viewing the lunar surface, at some appropriate period during a manned landing on the lunar surface, with a bank of proportional counters mounted in sector one of the service module of the Apollo spacecraft. In order to differentiate among the various fluorescent X-ray lines, and the primary X-radiation backscattered from the lunar surface, a balanced filter technique and pulse height analysis is employed. Background radiation induced by cosmic radiation on the detector walls is reduced electronically by pulse shape discrimination. The essential comparison of incident radiation to fluorescent radiation is done by continuously monitoring the incident solar flux, using a sun oriented monitor.

Instrument Description

The experimental arrangement to be mounted in the Apollo spacecraft consists of a sensor assembly which views the moon's surface, a solar monitor which measures the incoming solar X-ray flux, and an electronic interface between the various detectors and the spacecraft telemetry. The system will consist of three lunar oriented proportional counters with a total effective area of about 75 cm^2, and one solar oriented proportional counter with an area of about 3 mm^2. Two of the lunar counters will have filters in the field of view which is plus and minus 30 degrees square. These filters are to be used to supply differential spectra, and to improve energy resolution. Each of the lunar detectors will be made of beryllium with 0.001″ Be windows, and a P-10 gas filling at about 1 atmosphere.

Each lunar detector will feed into a charge sensitive preamplifier, an amplifier, and finally an 8 window stacked energy discriminator. The discriminator windows will drive a 16 bit scaler which will accumulate events for approximately 8 seconds, at which time the contents will be parallel shifted to telemetry readout registers (time shared by all four detectors) for readout to the spacecraft telemetry PCM. The proposed readout rate will be by a 10 pulse per second commutation technique.

Each of the three lunar stacked discriminator sets will be paralleled by a pulse shape discriminator to determine and reject high energy particle events detected in the counters.

A flow diagram of the electronic system is shown in Figure 3.14 and Figure 3.15 shows the experiment in exploded view. The system consists of a honeycomb collimator with a 60 deg. field of view, an assembly for holding the filters, a system of radioactive sources for calibration in flight, the detectors, and the electronics package.

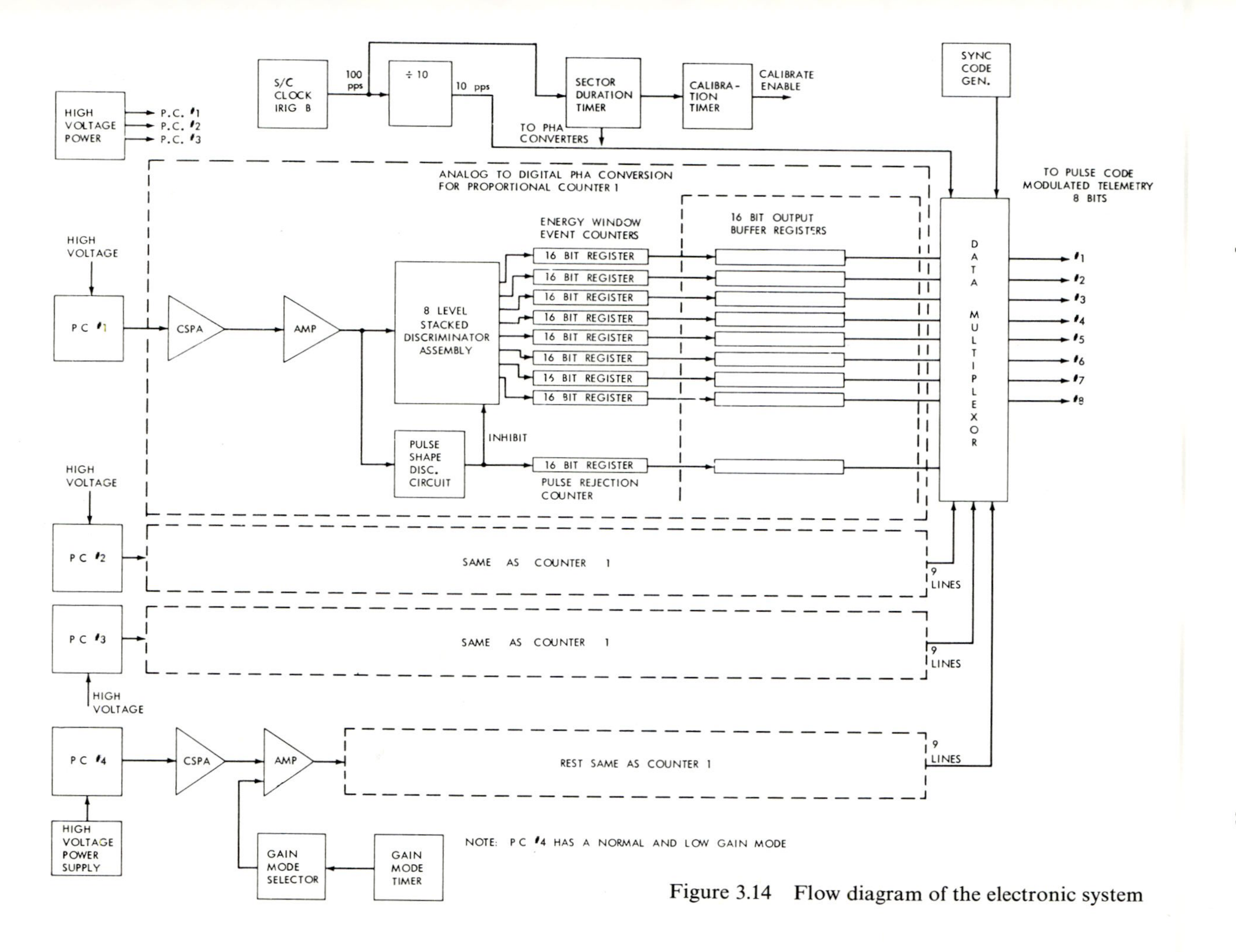

Figure 3.14 Flow diagram of the electronic system

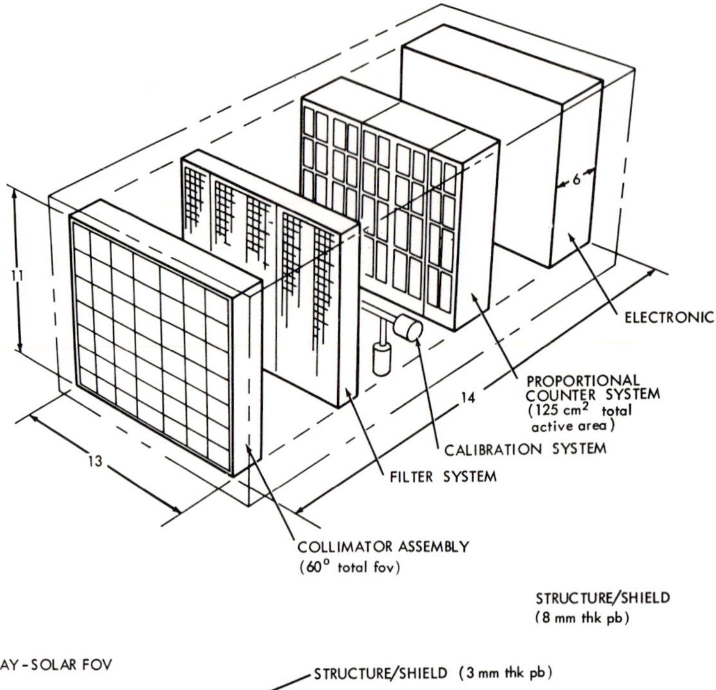

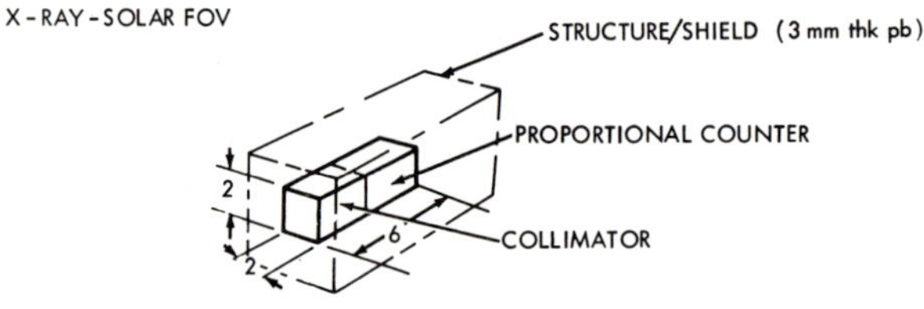

Figure 3.15 Experiment in exploded view

Orbital Analysis of Surface Composition by Gamma Ray Spectroscopy

The results of the Ranger and Luna flights are of considerable importance in the development of future gamma ray experiments. The Ranger and Luna experiments were discussed in detail in Chapter 2. A gamma ray spectrometer will be flown on Apollo 16 and 17. The experimenters group is made up of J. Arnold of the University of California, San Diego, as Principal Investigator, and A. Metzger of the Jet Propulsion Laboratory, L. Peterson of the University of California, San Diego, and J. Trombka of the Goddard Space Flight Center as co-investigators.

Experiment Approach

Details of the experimental method were considered in the previous chapter, thus only a brief discussion will be included here.

The gamma-ray spectrometer will measure the 0.3—10 Mev radiation flux from the surface of the moon while in orbit around the moon. This has two components: one is the decay of the naturally occurring radioactive elements in the lunar surface materials (potassium, uranium and thorium and their daughter products), and the other is produced by cosmic ray interactions with the nucleii of the chemical elements making up the lunar surface. The intensities of the natural gamma rays is a sensitive function of the degree of magmatic differentiation.

The observation of characteristic gamma ray lines produced by decay and nuclear interaction will be correlated with abundances and the spacecraft trajectory. Spatial resolution will be defined by the field of view to the horizon. Measurements will be made in cis-lunar space with the same geometry as the lunar measurements in order to obtain background reference data. In addition these cis-lunar measurements will provide the spectrum of the galactic gamma ray flux.

Equipment Description

Figure 3.16 is a schematic of the detector system to be used. The primary sensor is a large cylindrical NaI(*Tl*) crystal, approximately $2\frac{3}{4}''$ in diameter and length. The crystal is optically coupled to a ruggedized 3″ diameter photomultiplier tube.

Surrounding the inorganic scintillator on all sides, except the face to which the photomultiplier tube is coupled is a shell of plastic scintillator. The latter has poor efficiency for the detection of gamma rays above 0.1 mev, but responds effectively to charged particles. Operated in anticoincidence with the central detector, the plastic scintillator will practically eliminate any contribution from charged particles to the gamma ray spectrum. This is a modification in design from the Ranger experiment described in Chapter 2 where a phoswich arrangement was employed. Since NaI(*Tl*) and the plastic scintillator are separated by an opaque wall, the plastic scintillator requires its own photomultiplier tube. The detector assembly is mounted on the end of a twenty five foot boom in order to minimize effect of the background produced by cosmic ray induced gamma radiation in the Command and Service Module of the Apollo spacecraft.

The electronics subsystem includes amplifiers for the detector signals, a 512 channel analog to digital converter, programming logic to format

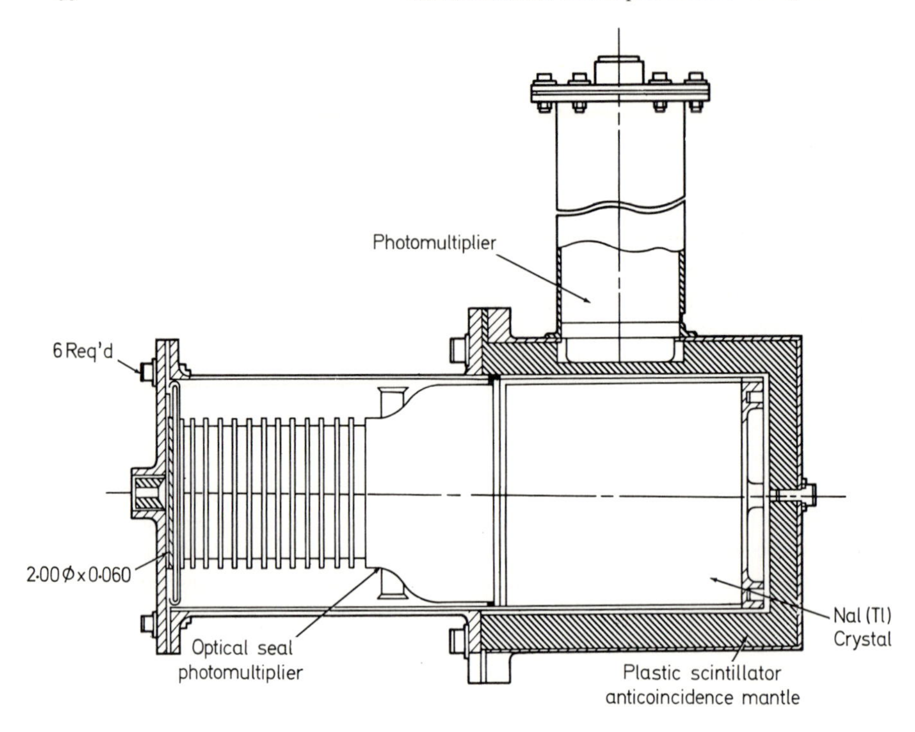

Figure 3.16 Proposed dual detector system. (After Metzger, 1964)

the data for presentation to the science data subsystem aboard the CSM, and a programmable high voltage supply for adjustment in case of instrumental gain drift. No memory accumulation is available for storing the pulse height spectra. The digitized pulse heights are transmitted in real time, and accumulation and sorting is done on the ground.

The analog to digital converter provides 511 channels of pulse height information over the energy range 0.3—10 mev. One channel will count the total number of pulses above the maximum digitized pulse. A reject pulse amplifier for the signal generated by the plastic scintillator will drive a threshold discriminator. Its output will provide anticoincidence action by closing the linear gate in the pulse height analysis channel.

Future Direction

A major factor in the ability to observe the presence of a discrete spectrum above the continuum is related to the detector resolution. The poorer the energy resolution, the more closely a discrete line will appear as a continuous distribution.

Solid state detectors such as lithium drifted germanium are receiving increasingly more attention in gamma ray spectroscopy. The major advantage, the great increase in resolution over the conventional NaI (*Tl*) crystal detector, leads to strongly enhanced signal to noise ratios and greater ease in the observation of the characteristic lines.

Before this type of detector can be used for future space flights, a number of problems (e.g., the need for constant cryoscopic cooling, and susceptibility to radiation damage) must be overcome. Efforts are underway to deal with these problems, and projections for future gamma ray experiments involve the use of solid state detectors.

Orbital Analysis of Soil Composition by Infrared Techniques

The possibility of employing infrared spectral analysis to obtain mineralogical information from lunar and planetary surfaces has been proposed by a number of investigators (Lyon, 1964; Hovis and Lowman, 1968; Aronson et al., 1966). An extensive evaluation of infrared techniques for lunar and planetary exploration was performed for NASA by Lyon at the Stanford Research Institute and published in 1965 as NASA Technical Note D-1871. The study included absorption studies of 370 rock and mineral samples, and reflection studies of 80 rocks. Spectral information was taken in the wavelength region of 2.5 to 25 microns, and emittance spectra were calculated from the reflectance data for some of the most important rock types.

In a subsequent investigation, Lyon (1965) published results for over 330 normal-emittance, reflectance, and transmittance spectra of roughened rock and mineral surfaces covering the range from 8 to 25 microns. Lyon drew the following conclusions:

1. Rock and mineral types can be determined from the wavelength position of the minima in the absorption or emission spectra. The details in the shape, intensity, and position of individual secondary maxima or minima can be used to infer differences between mineral groups and mineral assemblages in rocks.
2. Rock types can be defined if contrast is observed in the spectra.
3. Surface roughness, particle size, and physical discontinuities occurring within a few attenuation depths can have a marked effect on the infrared spectra of the material.
4. All the surfaces observed by emittance appear as gray bodies with an average emittance of 0.75 to 0.80 over the spectral range of 7.8 to 13 microns.
5. The effects of surface discontinuities are difficult to predict.
6. Absorption spectra have the highest spectral contrast and are the most definitive.

7. Where reflection methods are used, either flat polished surfaces must be prepared or hemispherical collection or irradiation techniques must be employed.

8. Emission analysis is the simplest to perform but the most complex to interpret. This type of measurement is most interfered with by surface porosity and particle size.

9. The emission method is most suited to field operations, particularly from orbit.

Aronson et al. (1966) proposed the use of the far infrared (15 to 500 microns) for the remote sensing of lunar and planetary surfaces. They attributed the neglect of this spectral region to the inherently low signal to noise ratio, and stated that the far infrared region is exceedingly rich in spectral information for silicate minerals. Hovis and Callahan (1966) showed that the reststrahlen (residual rays) of silicate-bearing minerals vary in frequency with the concentration of the silicate, and that the reststrahlen peaks vary from 8.5 to 11 microns depending on the types of rocks. Goetz and Westphal (1967) described a method for measuring spectral emissivity differences between two points on the lunar surface in the 8 to 13 micron region from a terrestrial observatory. The method they employed removes the basic atmospheric absorption, and allows the integration of many spectra in order to reduce the uncertainties produced by fluctuations in atmospheric and detector noise. Goetz (1968) published the results of a lunar survey involving some 22 lunar points using the spectral region of 8 to 13 microns. He concluded that most of the lunar regions surveyed appeared to be homogeneous in the 8 to 13 micron region. He attributed this to the roughening of the surface by micrometeorite bombardment or radiation effects. Two points, Plato and Mare Humorous, showed spectral contrasts due, perhaps, to compositional differences.

Principles

The principles which serve as the basis for infrared spectrophotometry, applied to the compositional analysis of lunar and planetary soils, have been described in detail by Lyon (1964, 1965). A brief review of Lyon's treatment follows:

Rock composition can be determined by measuring wavelength shifts in the fundamental Si—O vibration in the 9 to 11 micron band. For carefully prepared samples (well polished) the normal emittance spectrum "will permit the definition of individual minerals and/or glass, and the rock composition can be determined with fair precision".

The composition of rocks with rough or unprepared surfaces can also be determined using measurements of emittance by hemispherical spectral reflectance. However, the precision is less because of increased scatter. It has been observed, for example, that either technique applied to high silica glasses such as tektites or low silica stoney meteorites (chondrites) shows a compositionally dependent wavelength shift of up to 2 microns. Sample roughness does not affect the peak shifts but rather decreases the spectral contrast.

It has also been shown that certain surfaces emit radiation in such a manner that the radiative characteristics are completely specified if their absolute temperatures are known. These surfaces are ideal thermal radiators (black bodies), and emit a continuum.

The radiance of a black body source is given by the expression: $W_{BB} = \sigma T^4$, where W_{BB} is the power per unit area radiated into a hemisphere at a given temperature, T, and σ is the Stefan-Boltzmann constant. The emittance of any other surface at a given temperature is defined by the ratio W_S/W_B, e.g., the emissive power of the surface compared to a black body radiating at the same temperature and under the same conditions.

Lyon has shown that the same techniques used for reflectance and absorption spectra can be used to obtain emission spectra. Such spectra, calculated from the relationship, $(\varepsilon\lambda = 1 - \rho\lambda)$, are shown in Figure 3.17. $(\rho\lambda)$ is the reflectance obtained from the polished surfaces at ambient

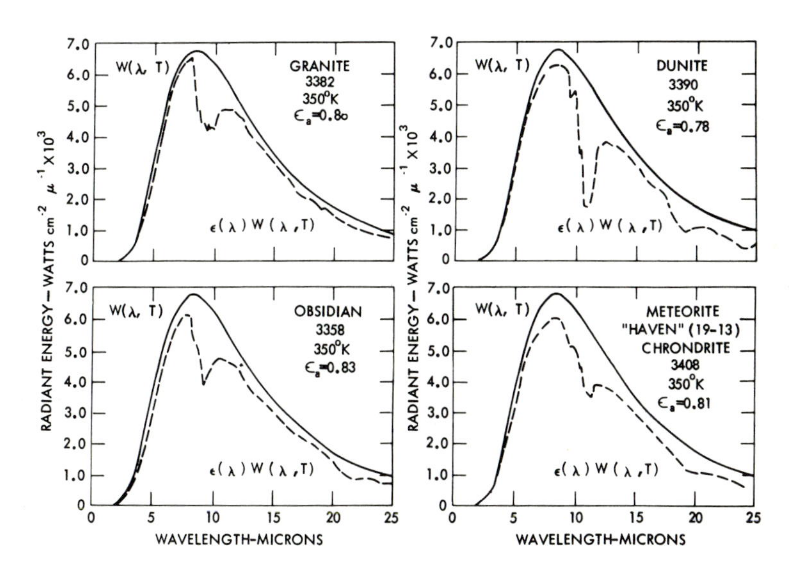

Figure 3.17 Emittance of granite, obsidian, dunite, and a stoney chondritic meteorite as a function of wavelength at 350 °K. Curves were calculated from reflectance data for the samples. (After Lyon, 1963)

temperatures. The spectra shown are for granite, dunite, obsidian, and a chondrite.

Lyon has conducted a series of studies on the normal spectral emittance of a number of solid rocks, powdered rocks, and rock forming

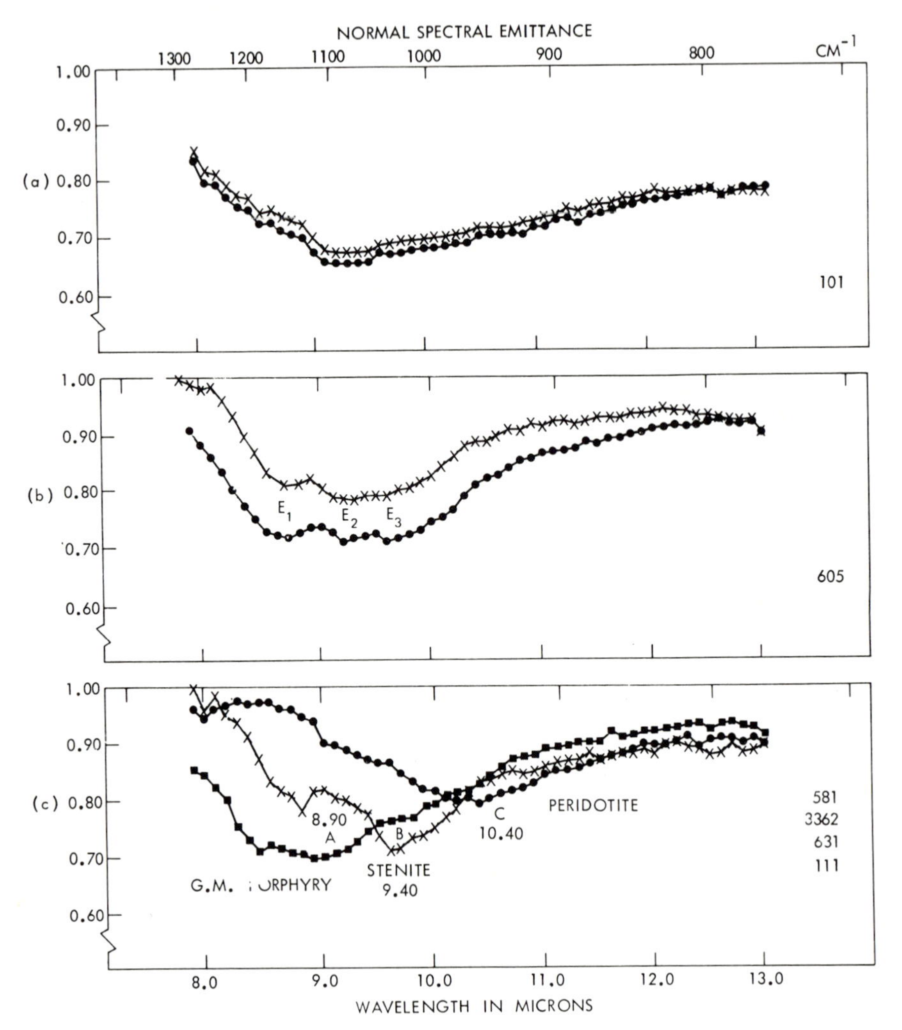

Figure 3.18 Normal spectral emittance curves. (a) Variance for each of two sawed faces using a quartz basalt sample, (b) normal spectral emittance curves for repeated measurements on a granite gneiss showing variance in measurements. Although the absolute levels of the two curves are different, the features E_1, E_2, and E_3 show in both, (c) normal spectral emittance curves for samples of quartz mononite porphory, syenite and peridotite. The spectral minima are displaced among these three compositions. (After Lyon, 1965)

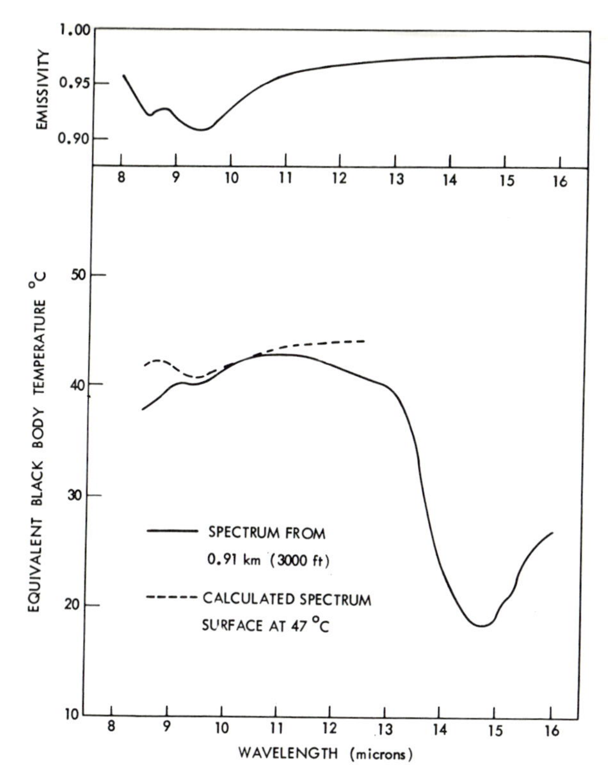

Figure 3.19 Thermal emission spectrum from Bristol Dry Lake, Calif. (After Hovis, 1968)

minerals. An attempt was made to duplicate field conditions by using rough, unbroken surfaces radiating into the spectrometer optics. Laboratory studies were performed with a single beam Perkin-Elmer PE-112, modified for direct emission measurements. The rocks were heated by means of a small water cooled furnace with an internal sample heater. Some of the results are given in Figure 3.18 which demonstrate repeated measurements on a quartz basalt, a granite gneiss, and a quartz monzonite. The figures show normal emittance curves for the different materials. Although the three samples have about the same average emittance, the spectral minima are markedly displaced for each of the different compositions.

Some infrared data has also been taken recently under flight conditions from an aircraft, and described by Hovis (1968). Figures 3.19 and 3.20 show some examples of the results obtained. These are reststrahlen spectra from different formations such as a lava bed from a crater in California and a dry lake bed (Bristol Dry Lake, California).

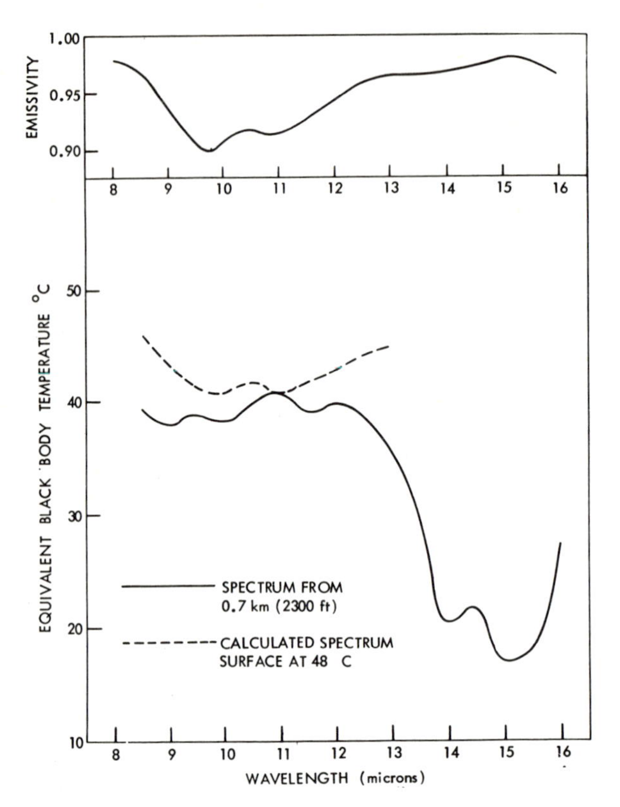

Figure 3.20 Thermal emission spectrum of Lava, Amboy Crater, Calif. (After Hovis, 1968)

Goddard Space Flight Center Four Channel Radiometer (Hovis and Lowman, 1968)

An example of an infrared experiment for compositional analysis is that proposed by Hovis and Lowman for use in the 1971 mission to Mars. The objectives for this experiment have been summarized as follows:

To provide a continuous record in segments of about a kilometer of the dominant rock composition in those areas covered by television display. The experiment should differentiate among the various rock types such as the ultrabasic, intermediate basic, and acidic rocks. The spatial resolution planned should provide data about the major physiographic units, central peaks and crater rims, and across structural lineaments. It is proposed that a determination of the rock types, combined with a study of the television pictures, will provide some possible answers to some of the major questions about the evolution of Mars.

A schematic representation of the flight instrument is shown in Figure 3.21. The instrument has been designed to resolve 0.5 milliradians spatially in the 8 to 11 micron region. The telescope is a Cassegranian with a primary mirror 8 inches in diameter, and a secondary obscuration of 0.4. Since the available data rates do not permit scanning, except for that provided by the moving spacecraft, the radiometer is to be fixed and boresighted with the television camera, scanning a strip one kilometer wide and resolving one kilometer elements along the strip at closest approach. Four channels will be used with appropriate interference filters centered at 9.0, 9.9, 10.7, and 11.5 microns. The first three wavelengths correspond to acidic, basic, and ultrabasic rocks, respectively. The 11.5 micron channel is for the spectral region where minerals are blackest, in order to provide true surface temperatures. Conceptually, the equivalent blackbody temperatures are measured simultaneously in all four channels on the same spot on the surface. Thus, a cooler temperature than the reference can only represent an emissivity effect because, as Hovis states, an object cannot have two temperatures at the same time. If two or three short wavelength channels appear cooler, it will indicate a mix of surface material and the degree of mixing.

A chopper is used to allow a view of deep space as a reference for both sets of detectors. This chopper, which is gold coated on both sides, acts

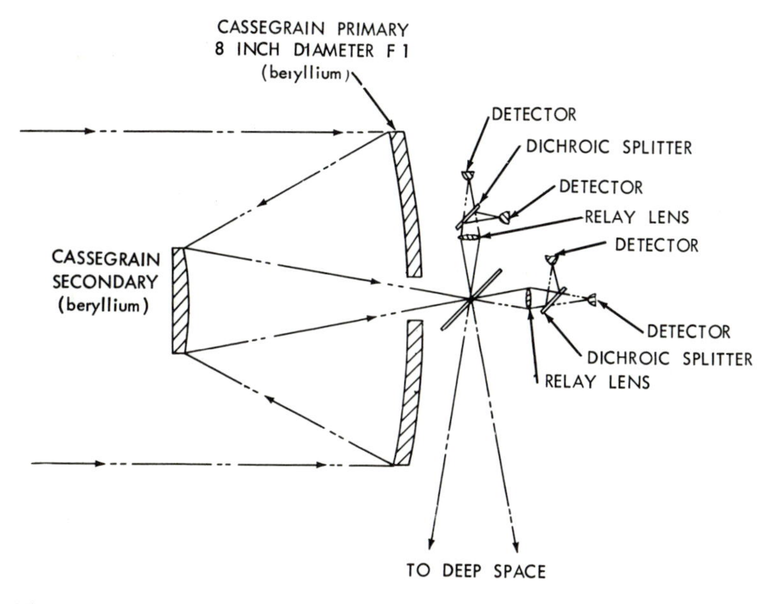

Figure 3.21 Optical schematic of proposed four-channel radiometer. (After Hovis, 1968)

as a beam splitter allowing one set of detectors to see the target reflected from the chopper and deep space. The reverse situation is true for the other detector pair.

As outlined, the total power consumption will be of the order of 9 watts and the experimental package will weigh approximately 15 lbs.

Orbital Analysis of the Lunar Surface by Alpha Particle Spectroscopy

We have discussed above two of the three elemental analysis experiments to be included aboard the Apollo 16 and 17 flights. The third experiment we shall describe, utilizes surface barrier detectors in order to determine the differential energy spectrum of alpha particle emanations from the lunar surface. The principal investigator for this experiment is P. Gorenstein of American Science and Engineering, Cambridge, Mass. and the coinvestigator is H. Gursky of the same institution.

Scientific Description

The alpha particle spectrometer will measure the rate of radon (^{220}Rn and ^{222}Rn) evolution from the moon by detecting the alpha particles that are produced in their decay. Decay of radon above the surface will result in the production of mono-energetic alpha particles plus an active deposit on the lunar surface that will emit additional alpha particles, and gamma radiation. The longer lived isotope ^{222}Rn ($T_{1/2}=3.8$ days) should have sufficient time to establish a uniform atmosphere of gravitationally trapped gas over the entire lunar surface. ^{220}Rn ($T_{1/2}=55$ sec) will undergo decay within a short distance of its point of evolution. The absence of any material around the moon that could slow the alpha particles allows a unique and sensitive way of detecting the radon evolution. The measurements will determine the gross rate of radon evolution from the moon by observation of the longer lived isotope ^{222}Rn and, will possibly detect localized sources of enhanced emission of radon by observing the decay of the shorter lived isotope ^{220}Rn.

There are several reasons for wishing to study radon evolution rates from the moon. One, the gross rate of evolution should be a function of the average concentration of uranium in the first meter of lunar material and its radon diffusion characteristics. With information from the gamma sensor, the concentration of uranium can be determined so that it is possible to estimate the diffusion characteristics of the surface.

In turn, the diffusion properties are related to the porosity and quantities of absorbed gases in the lunar surface.

Secondly, if there is significant diffusion of radon to the surface, then the active deposit from radon decay will increase the gamma ray activity of the surface. Hence, the alpha measurement is needed in order to subtract the effect of surface deposits and thus obtain a more meaningful interpretation of the gamma ray measurements of the uranium concentration.

Finally, the locations of regions of enhanced radon emission can be an indication of one or more of the following interesting features; the presence of crevices or fissures on the lunar surface, areas which release volatiles (e.g. volcanism), or possibly regions with unusual concentrations of thorium.

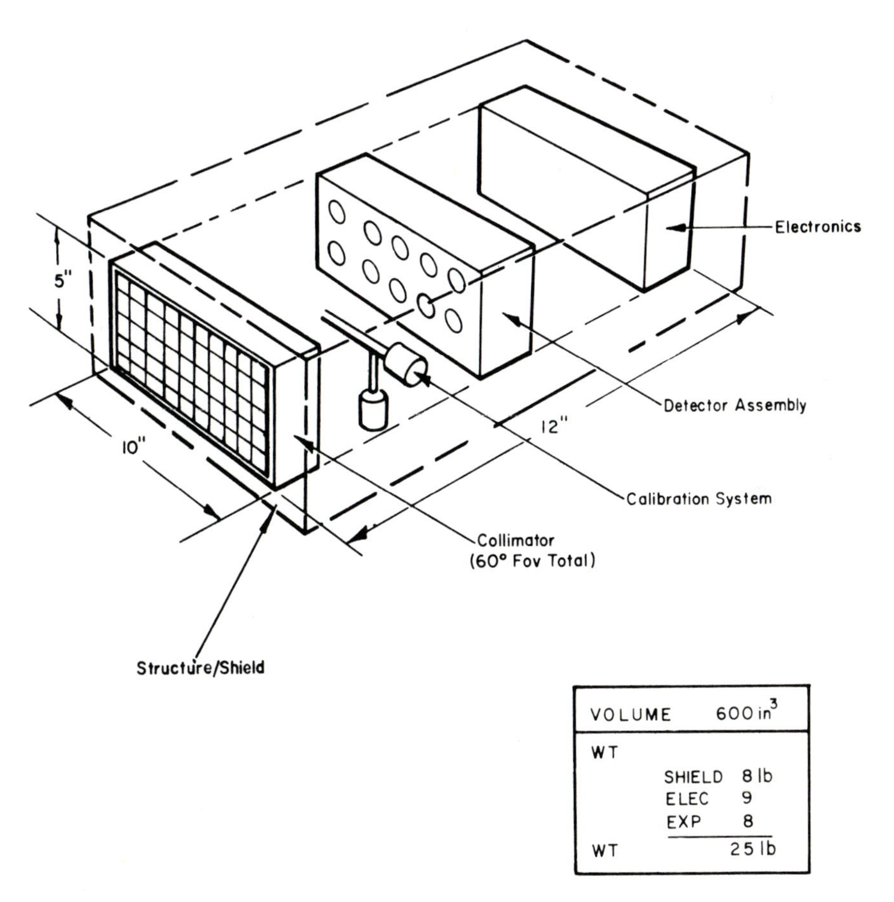

Figure 3.22 Schematic Illustration of Alpha-Particle Spectrometer

Instrument Description

Thirty square centimeters of active detector area will be divided among individually biased solid state silicon, surface barrier detectors. Figure 3.22 is a schematic representation of the instrument. Ten detectors will be used and the pulse height analyzer will have 256 channels.

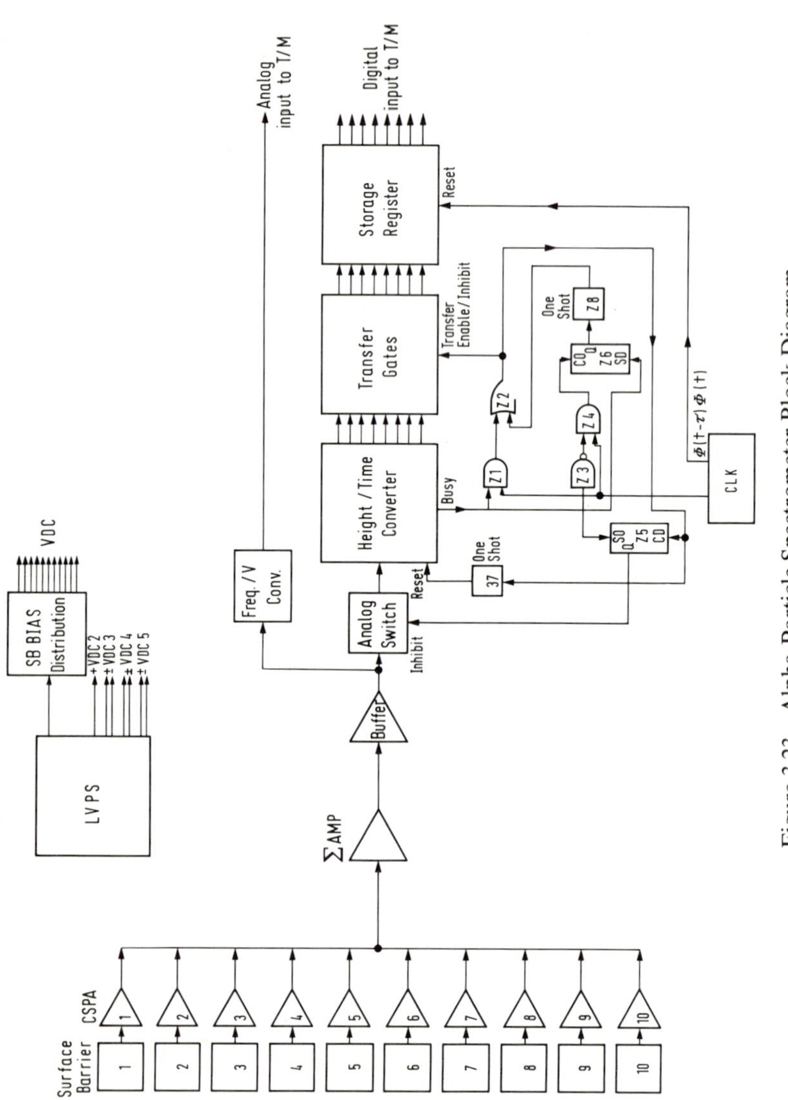

Figure 3.23 Alpha-Particle Spectrometer Block Diagram

A block diagram of the electronics is shown in Figure 3.23. The signal from each of the surface barrier detectors will be coupled into its respective charge sensitive preamplifier. The output of each preamplifier-detector combination will be adjusted so that for a given particle energy all outputs, from any of the ten detector-preamplifier combinations will appear identical. These signals will then be summed, amplified, and buffered to the processing level required by the analog to digital converter (256 channels). The eight parallel line output of the analog to digital converter will be transferred through gates into a storage register. The output of the storage register will then be sampled by telemetry with channel by channel being transmitted to the earth. The data will be accumulated and sorted into pulse height spectra on the earth similar to the method employed for the gamma ray experiment.

Neutron Methods

Neutron induced excitation methods for the chemical analyses of lunar and planetary surfaces have been proposed by a number of investigators early in the exploration program (Schrader et al., 1961; Metzger et al., 1962; Schrader et al., 1962; Lee et al., 1962; Monaghan et al., 1963; Trombka et al., 1963; and Greenwood et al., 1963). Included was the proposal to use both natural and accelerator neutron sources. In 1964, NASA organized a team of investigators to develop the most reasonable neutron methods and to implement the necessary instrumentation. A preliminary report by the members of this group was represented in Science (Caldwell et al. 1966). The emphasis in this report was on the use of a machine neutron source. The type of source and the instrumental method was shown to depend strongly on the nature and extent of the mission being planned. A description of the various mechanisms and instrumentation for performing elemental analysis are outlined below.

Principles of the Neutron-Gamma Method

Elemental analysis can be obtained from the neutron irradiation of target material by utilizing the spectral and/or the temporal characteristics of the gamma rays produced from the neutron-matter interactions. The emitted radiation can be classified as prompt, capture, and activation gamma rays.

The interaction of neutrons with matter depends on their energies and the nuclear species present. With the exception of slow neutrons, the interaction is unaffected by the chemical or physical state of matter. Fast neutrons (i.e., $E > 10$ MeV) interact with nuclei either by scattering or by nuclear reaction. Scattering may be either elastic with the neutron's kinetic energy conserved (see Figure 3.24), or inelastic with some part of the neutron energy transferred to the nucleus. The subsequent de-excitation by gamma ray emission is shown in Figure 3.25. Some typical neutron-nuclei reactions involve neutron capture followed by the emission of alpha particles, gamma rays, protons, or neutrons. For low to medium mass nuclei, which are the major constituents of common rocks, the most probable interactions are elastic, inelastic, and $(n, 2n)$. The inelastic cross section for the major rock forming elements (Si, O, Fe, Mg, Al, Na, K, Ca) are approximately the same.

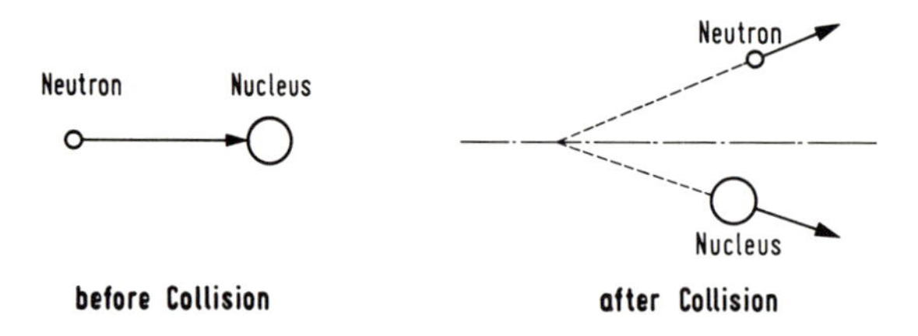

Figure 3.24 Neutron Elastic Scattering

At lower neutron energies (of the order of several hundred kev), the neutron reaction and inelastic scattering cross sections become very small. For the light nuclei, these cross sections are small for energies even up to the order of several MeV. The notable exceptions are ^{6}Li and ^{10}B. The remaining interaction which does occur is neutron capture (see Figure 3.26). The neutron remains in the nucleus, and the excess energy is emitted in the form of gamma rays by a process named radiative capture. The new isotope formed by this capture may be either stable or unstable. The unstable nuclei decay with the emission of an electron or positron, or by nuclear electron capture, along with a neutrino emission at a characteristic rate or lifetime. This process of producing unstable nuclei is called activation (see Figure 3.27). The resulting nuclei, if left in an excited state by the decay, will de-excite by the emission of a characteristic gamma ray.

Experiment Configuration and Operation

The neutron-gamma technique for elemental analysis involves the excitation of a surface by fast neutron bombardment. The gamma rays produced by the above-mentioned processes of inelastic scattering, capture, and activation are emitted in different time domains relative to the time of production of the fast neutron by the source. It is possible to

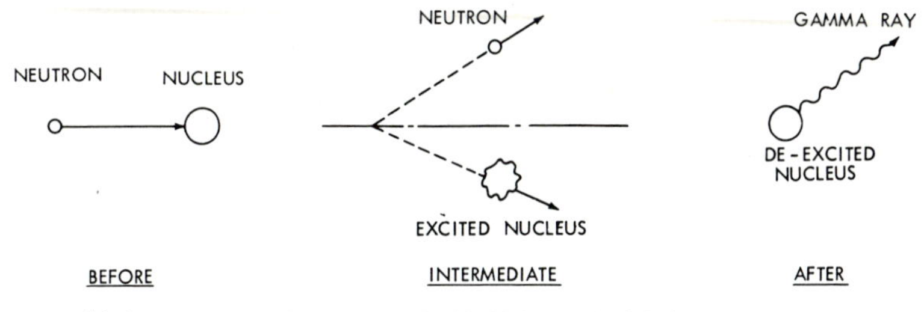

(1) GAMMA RAY IS CHARACTERISTIC OF ORIGINAL NUCLEUS
(2) THRESHOLD FOR PROCESS VARIES; TYPICALLY 0.5-5 MEV
(3) TIME FOR PROCESS $\sim 10^{-14}$ SEC

Figure 3.25 Inelastic scattering

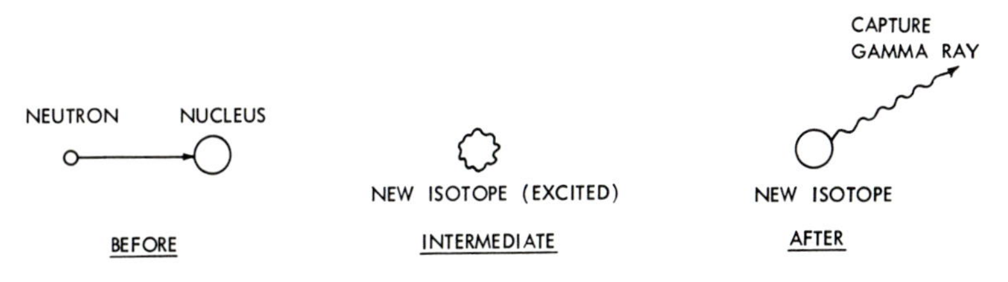

(1) GAMMA RAY IS CHARACTERISTIC OF NEW ISOTOPE
(2) PROCESS IS MOST LIKELY TO OCCUR AT LOW NEUTRON ENERGIES
(3) TIME FOR PROCESS $\sim 10^{-14}$ SEC

Figure 3.26 Radiative capture

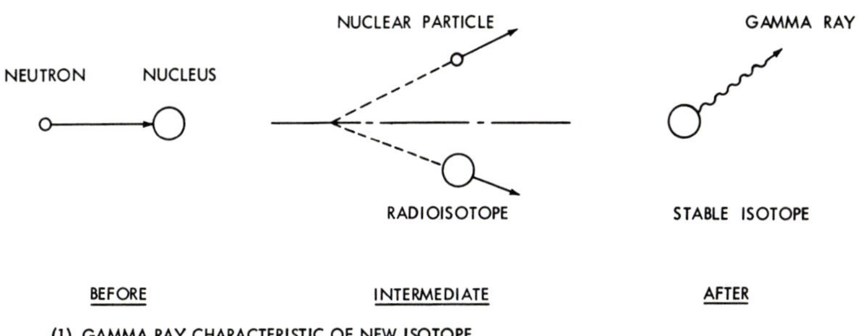

(1) GAMMA RAY CHARACTERISTIC OF NEW ISOTOPE
(2) PROCESS OCCURS WITH FAST NEUTRONS
(3) TIME FOR PROCESS TYPICALLY RANGES FROM SECONDS TO HOURS

Figure 3.27 Activation

separate the variously produced gamma rays by time discrimination. The use of a pulsed source of neutrons and time gated spectral measurements provides a convenient means of separation. The resulting spectra can be used to identify many elements of interest as shown in Figure 3.28. Discrete gamma ray spectral lines are shown for oxygen, magnesium, aluminum, silicon, and iron. The intensities of different lines for a given element and a given production mechanism. are normalized for comparison (Caldwell et al., 1966). Not included in the figure are hydrogen (capture gamma at 2.2 MeV) and carbon (inelastic gamma at 4.4 MeV).

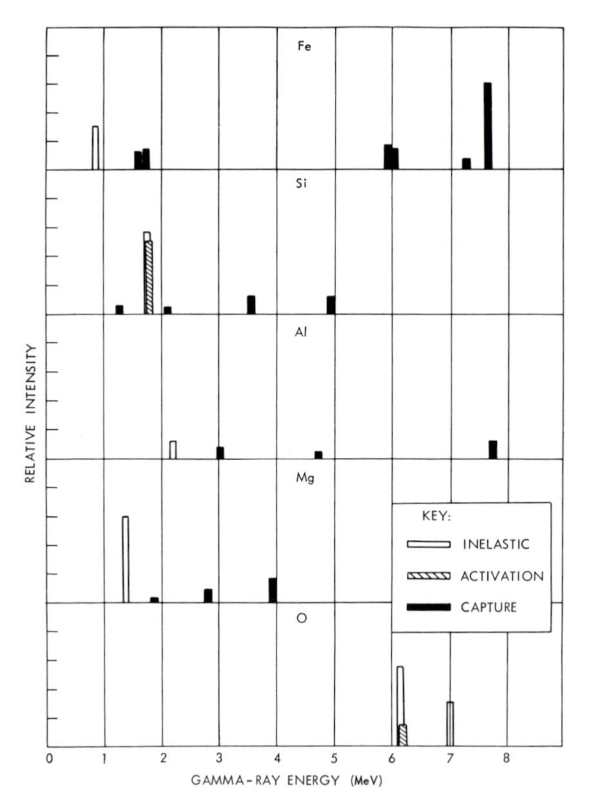

Figure 3.28 Ideal gamma ray spectrum induced by 14 MeV pulsed neutron source

A portable neutron generator has been developed for space flight application capable of 10^4—10^6 neutrons per pulse at the source, at a repetition rate of 500 to 1500 pulses per second. Using such a source, with proper attention to the time dependent phenomena, it should be possible to readily identify these elements: hydrogen, carbon, oxygen, silicon, aluminum, magnesium, iron, titanium, potassium, thorium, and uranium. The last three elements would be identified in the quiescent spectrum, obtained prior to the turning on of the neutron source.

One additional measurement which can be made is the neutron "die away". After a neutron burst, the rate at which the flux of both the "epi-thermal" neutrons (i.e. neutrons between 1 ev to 100 ev) and the "thermal" neutrons (less than 1 ev) decreases can be related to the hydrogen content of the soil. This rate of flux decrease is the so-called "die away". The mean life time, λ, of the thermal neutrons in a given media depends on two processes: absorption of the neutrons, and the leakage of the thermal neutrons out of the surface. An approximate formula for λ is:
$\lambda = v\Sigma_a + DB^2$, where v is the thermal neutron velocity, Σ_a is the macroscopic absorption cross section, and B^2 is the effective "Buckling", a measure of the geometric configuration and, for the application under consideration, a constant for all practical purposes. D and Σ_a can be approximated as follows: $D^{-1} = \rho \Sigma W_i$ (β scattering)$_i$, $\Sigma_a = \rho \Sigma W_i$ (β capture)$_i$, where ρ = bulk density, W_i = weight fraction of ith type element, and β scattering and β capture are related to the macroscopic transport and capture cross section. λ is determined in practice by a measure of the temporal behavior (i.e., the "die away") of the thermal flux. The weight fractions of the various elements can be determined from the gamma ray spectra, and the bulk density can then be determined.

If decay of the epithermal neutron flux can be observed, the decay constant should be dependent on hydrogen content. The epithermal decay constant, λ, is related to the quantity,

$$\sum_i \frac{W_i \xi_i \sigma_{si}}{A_i},$$

where ξ_i and σ_{si} are, respectively, the average logarithmic decrement per collision and the average scattering cross section for fast neutrons for the ith type nucleus.

Even though the amount of hydrogen present may be small, its presence will greatly effect ξ_i and thus effect the value of λ. Thus an experimentally determined λ epithermal will indicate either the amount or change in amount of hydrogen present.

Combined Neutron Gamma Experiment

The combined pulsed neutron experiment can be most expeditiously carried out in two modes depending on the operation of the neutron generator. One mode is characterized by a high pulse repetition rate-low flux per burst. The second mode is a low repetition rate-high flux per burst, Figure 3.29 shows the geometrical configuration of the neutron system, including an epithermal neutron detector for the epi-thermal die away measurements. The nuclear phenomena shown are those appropriate to the high repetition rate mode.

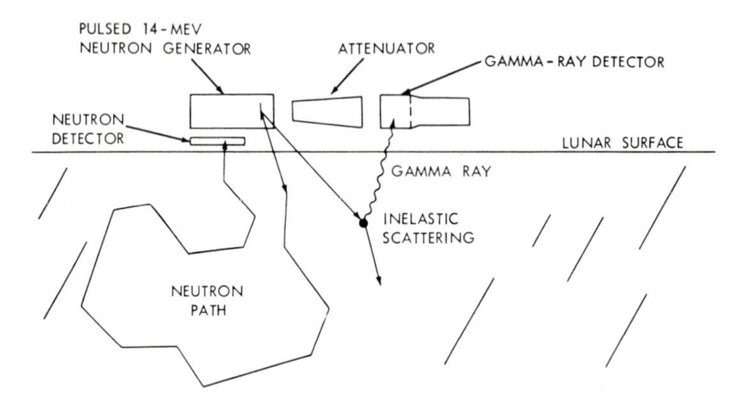

Figure 3.29 Geometrical configuration for combined pulsed neutron experiment

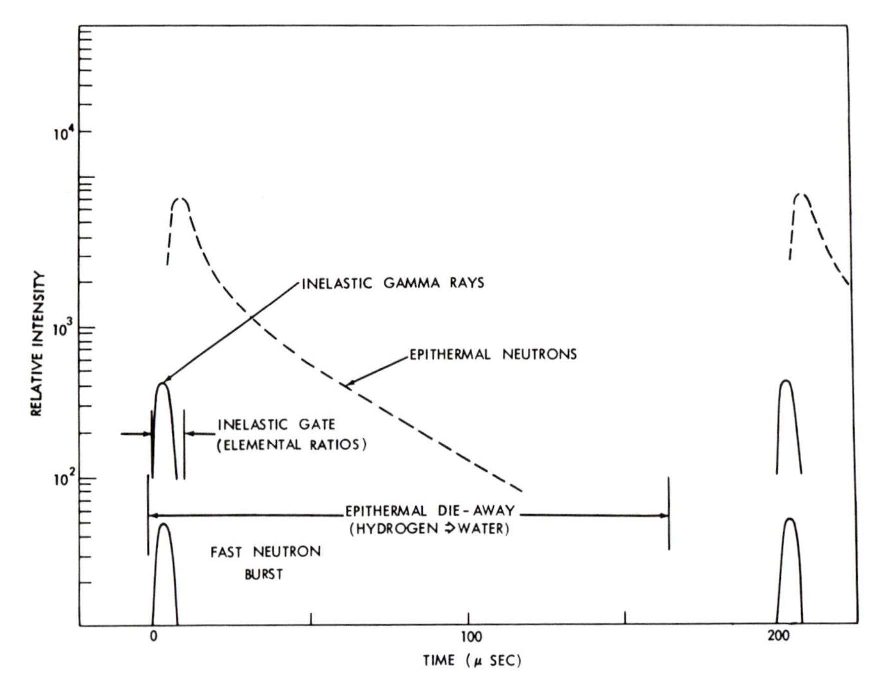

Figure 3.30 Timing diagram for combined pulsed neutron experiment. High repetition rate

Figure 3.30 (Mills et al., 1969) shows a timing diagram for the high repetition rate. The ordinate is the relative intensity plotted on a logarithmic scale, the abscissa is a linear time axis. The fast neutron burst from the 14 MeV generator is shown in the shaded area. In this illustration, a short production burst with a 200 μsec period is assumed.

The gamma rays which are produced coincidentally with the burst of neutrons result from inelastic scattering. The inelastic gamma ray production begins and ends sharply with the onset and termination of the fast neutron burst.

The counting rate observed by the epithermal neutron detector will look qualitatively like the dashed curve. The epithermal neutron intensity builds up during irradiation, and dies away following each neutron burst.

The timing diagram for the low repetition rate operation of the neutron source is shown in Figure 3.31 (Mills et al., 1969). The fast neutron bursts are still 5 μsec in duration, but they are separated by 200 μsec. The total number of neutrons per burst is roughly ten times greater than in the high repetition rate mode.

Within a hundred or so μsec after each neutron burst, practically all the neutrons in the sample are thermalized. The thermal neutron density decreases, or dies away, with time due to capture and leakage out of the surface. Because the capture gamma ray intensity is directly related to the thermal neutron density, the "die away" has a time dependence which is qualitatively as shown in Figure 3.30.

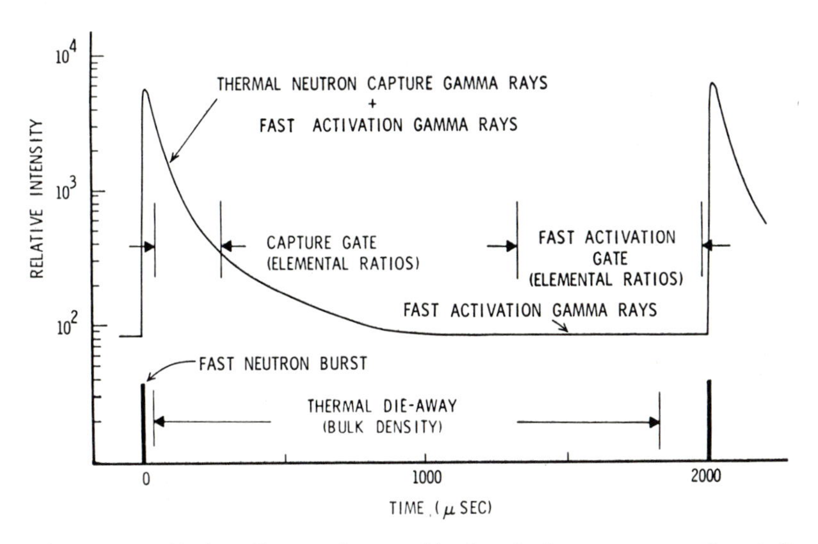

Figure 3.31 Timing diagram for combined pulsed neutron experiment. Low repetition rate

The radiative capture gamma ray are measured during the die away period. However, the gamma ray intensity does not drop to zero in the time between the end of the capture gamma ray-die away and the following neutron burst. Radiation from the naturally occurring long lived isotopes is present throughout the source burst cycle. In addition, the radioactive isotopes produced in the surface material from (n, α), (n, p), and $(n, 2n)$ fast neutron activation emit gamma rays throughout the total time period. If the intensity of all the gamma rays is measured during the interval labelled "Thermal die-away" (as shown in Figure 3.31), a curve consisting of only capture intensity as a function of time can be obtained. The bulk density of the surface may be inferred.

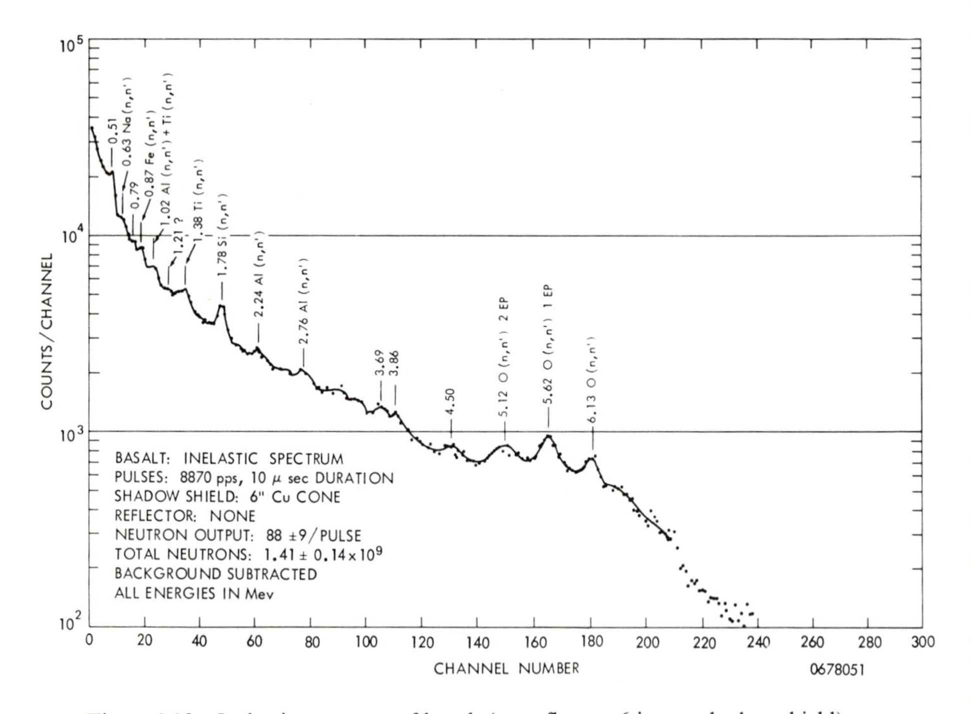

Figure 3.32 Inelastic spectrum of basalt (no reflector, 6 in. cu. shadow shield)

The experiments described by the timing diagram shown in Figures 3.30 and 3.31 can be performed by using a programmed time sequencing control system. Some typical results obtained by such measurements are shown in Figures 3.32 to 3.35. For the details of analysis see the cited references.

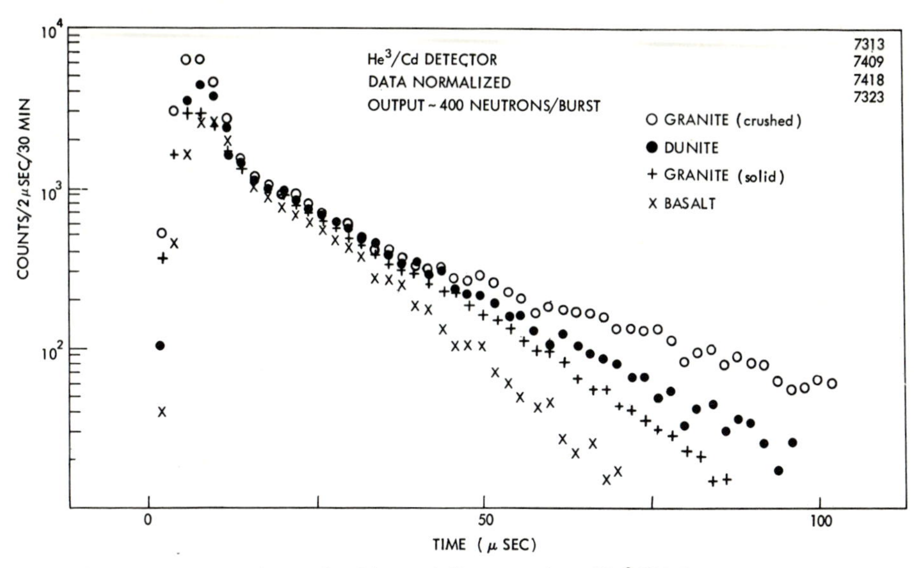

Figure 3.33 Experimental epithermal die-away data. He^3/Cd detector

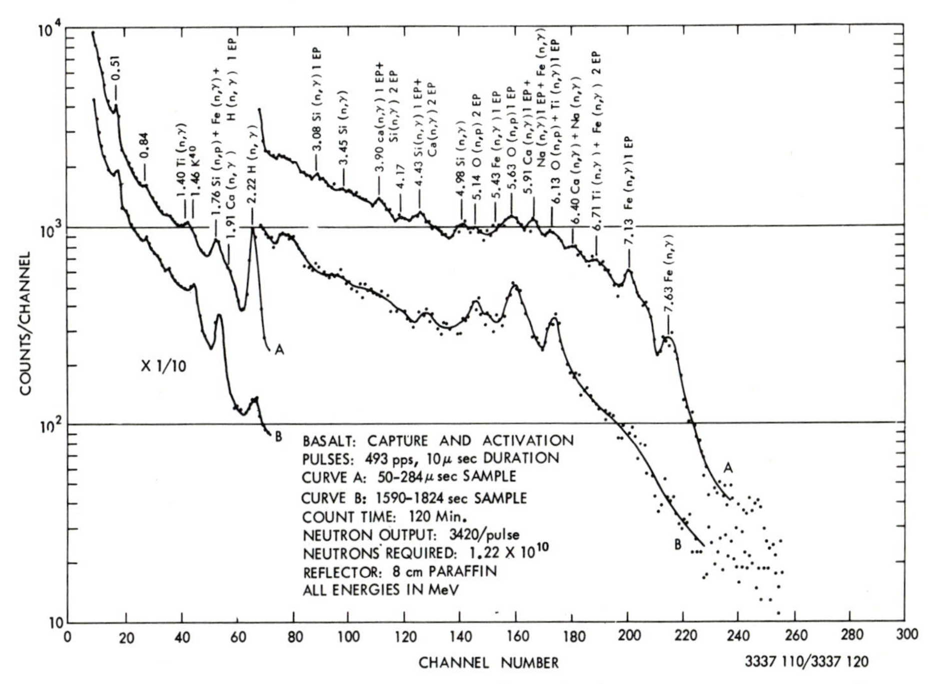

Figure 3.34 High energy capture and cyclic activation gamma-ray spectra of basalt (3420 neutrons per pulse)

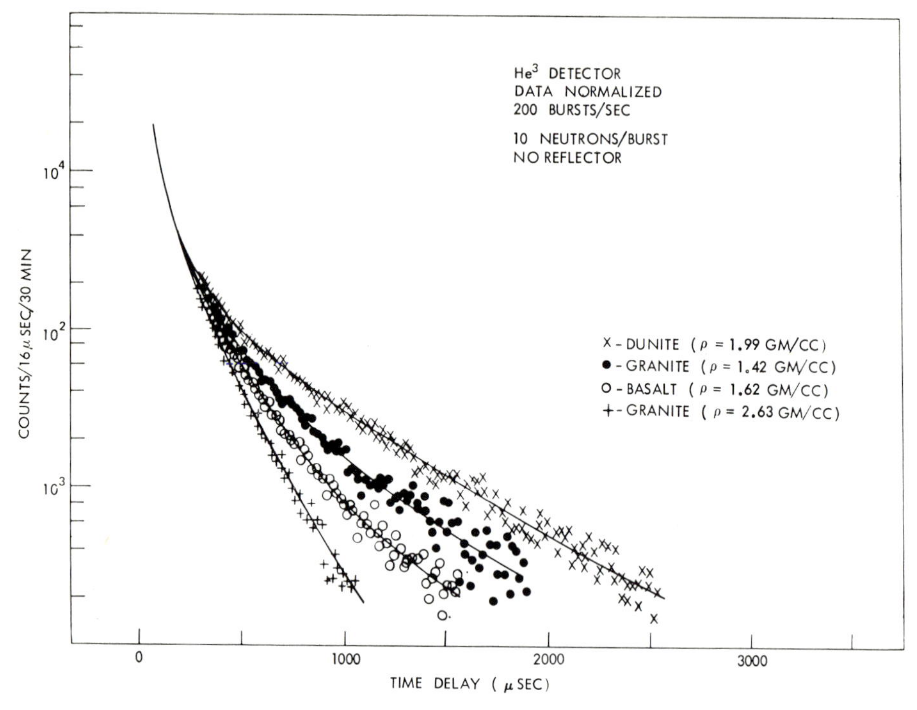

Figure 3.35 Experimental thermal neutron die-away data. He^3 detector

Radioisotopic Neutron Sources

The recent development of intensly emitting radioisotopic neutron sources has made it possible to design simpler and more reliable experiments for neutron-gamma studies. The device to be described is tailored to perform elemental analysis as quickly as possible to aid in such functions as sample selection and geochemical mapping. It avoids the inevitable delays and long counting times associated with activation analysis techniques. The technique is built around the use of ^{252}Cf, an exceptionally strong neutron emitter, and involves spectral measurements during the irradiation cycle. The measured spectra contain information on radiative capture gamma rays as well as gamma rays emitted by some short lived, activated species.

Instrumentation built around a ^{252}Cf source has a number of advantages:

1. It is possible to use relatively small sources of a ^{252}Cf neutron source ($\sim 10\,\mu$gm) which in turn requires a small shield (~ 10 lbs).

2. The source itself emits no significant gamma ray flux to interfere with the measurements.

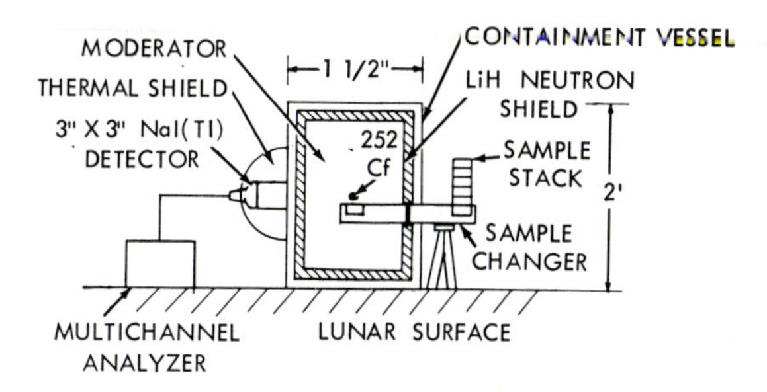

(a) MULTI-SAMPLE EXPERIMENT

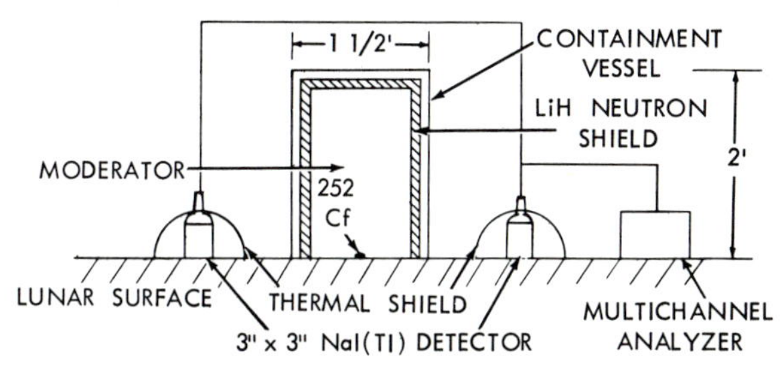

(b) IN SITU EXPERIMENT

Figure 3.36 Active gamma ray experiment. (a) Multi-sample experiment, (b) in situ experiment

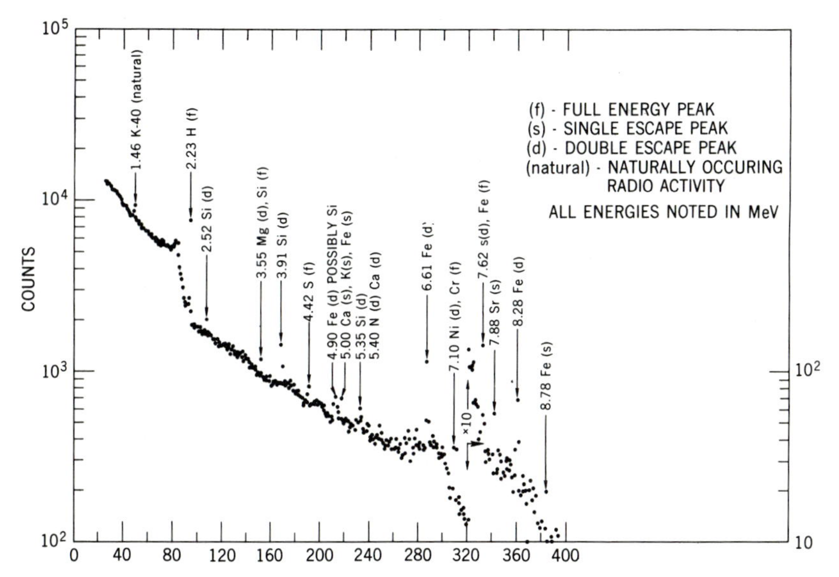

Figure 3.37 Prompt capture and activation gamma pulse height spectrum using ^{252}Cf source and solid state detector (35 cm^2 Ge(Li))

3. The radiative capture gamma rays have energies up to 10 MeV, and are capable of penetrating through the shield thus permitting simultaneous irradiation and measurement.

Three possible configurations for a possible lunar and planetary exploration are under consideration (Trombka et al., 1969).

1. Multi-Sample Analysis Facility: $A\,10\,\mu g$ ^{252}Cf source ($\sim 10^7$ neutrons/sec) is used with a $3'' \times 3''$ NaI(Tl) detector as shown in Figure 3.36a. The neutron source is mounted in a deuterated hydrocarbon shield (paraffin) with a LiH outer shield. The samples are introduced into the shield, near the source, and the gamma ray measurements made during the irradiation cycle.

2. In situ Measurements (Figure 3.36): In this mode, the source and shield are to be placed directly on the surface to be measured and the source lowered to contact the surface. Single or multiple detectors are

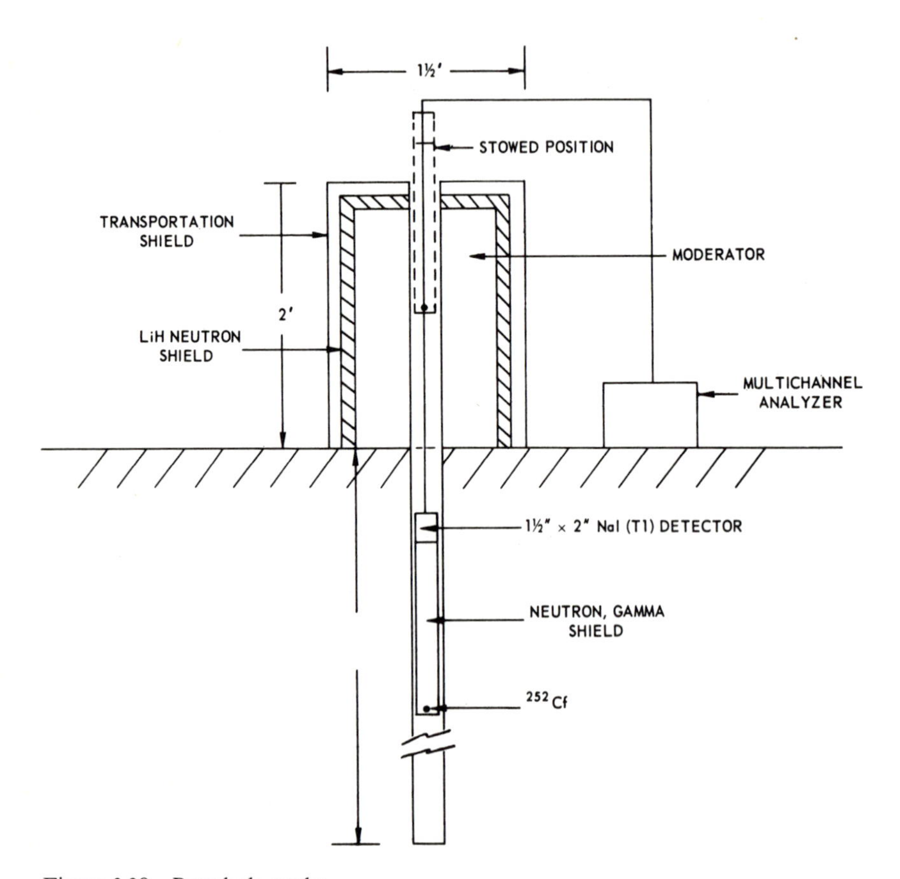

Figure 3.38 Bore hole probe

placed on the surface away from the shield which leads to spectral measurements from a relatively large sample. Such results are useful for mapping. An example of this kind of measurement, using a Ge(Li) type detector, is shown in Figure 3.37.

3. Down Hole Probe: Figure 3.38 shows a third option for performing compositional profiling down a bore hole. The technique would involve slowly lowering the source and detector in a stepwise manner while making gamma ray measurements at each step.

In summary, studies have shown that neutron methods show great promise for elemental analysis of surfaces. Both isotopic and accelerator sources can be used. Only a few possible configurations have been discussed, however the instrumentation is sufficiently versatile that other configurations are possible to extend further the usefulness of the techniques. The particular configuration that is chosen will strongly depend on the particular space mission and goals.

Mass Spectroscopy for Lunar and Planetary Exploration

The fundamental capability of mass spectroscopy to provide analytical information about planetary surfaces and atmospheres has made it a promising technique for space exploration. Mass spectrometers for the determination of upper atmosphere constituents have already been successfully flown on sounding rockets and satellites (Spencer and Reber, 1962). Kendall et al., in a study performed for Marshall Space Flight Center (1964), concluded that the most promising type of equipment for use in lunar exploration appears to be a monopole mass analyzer with an electron bombardment ion source and electron multiplier detector. In a similar sort of study performed for the Goddard Space Flight Center, Kreismann (1969) evaluated a solids mass spectrometer using a sputter-ion source and a compact double focusing mass spectrometer. Hoffman (1969) has proposed a small magnetic mass spectrometer, as part of a science payload to be carried aboard a lunar orbiter, to measure the high altitude gas atmosphere around the moon, and a similar device to measure gases on the lunar surface.

In principle, charged atoms and molecules are produced by electron or charged ion bombardment of a target. A beam of positive ions is drawn from the bombarded area, and the ions are sorted out according to their masses by a magnetic field. The separated masses are then displayed either photographically, as in a mass spectrograph, or detected electronically by various detectors as in a mass spectrometer. The nature of the spectrometer selected for flight experiments depends on a number of factors: the mass resolution required, the kinds of material being sampled, the initial spread of energies, memory effects, the mission profile, etc. It is obvious that the topic is extensive, and only a brief discussion will be given here describing representative instrumentation.

Mass Spectrometer for Atmospheric Gases (Spencer and Reber, 1962)

Figure 3.39 is a cross-section view of a double focusing mass spectrometer flown aboard an Explorer XVII satellite (1963) for quantitatively determining the atmospheric components at high altitudes. The purpose of the experiment was to obtain density values for helium, nitrogen, oxygen, and water; and to obtain more comprehensive data about the altitudinal variations and geographic, diurnal, and seasonal variations.

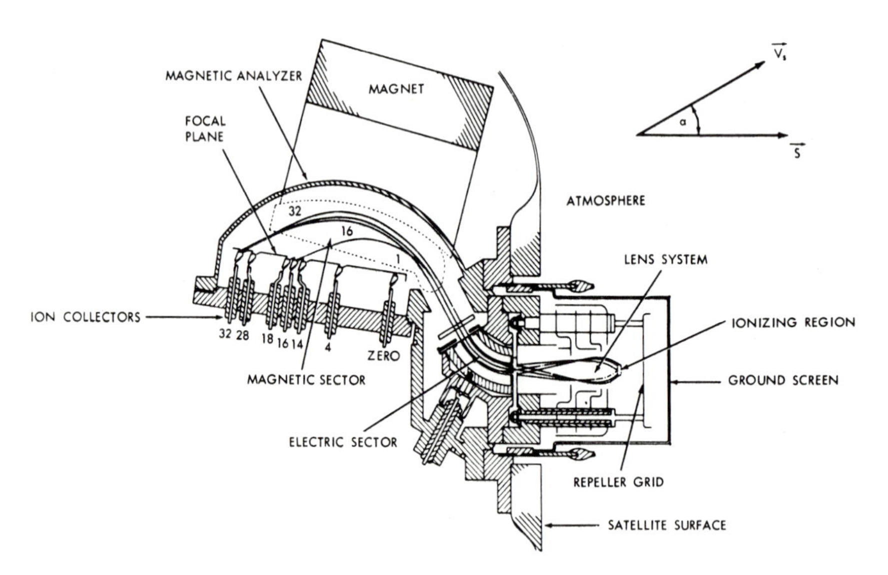

Figure 3.39 The mass spectrometer cross-section in the $x-y$ plane

The instrument shown is a tandem double-focusing instrument with a 60° electric sector and a 90° magnetic sector. The ion source and electrostatic lens preceding the mass analyzer were designed especially for use aboard a satellite. The tube itself is evacuated prior to and during launch, and the outer cover, referred to as a "break-off hat", is ejected on reaching a region of suitably low pressure.

The design philosophy of the mass spectrometer is built around the fact that the thermal velocities of the ambient particles are small relative to the satellite velocity of about 8 km/sec. Thus the velocity of the particles may be considered nearly constant and equal to the satellite velocity. In view of this, the apparent kinetic energy, $mv^2/2$, of the mass species is essentially proportional to their masses. For the heaviest mass sampled, molecular oxygen, the energy is of the order of 12 ev.

In the Explorer XVII mission, the mass spectrometer was located on the spin axis of the spacecraft. Thus, the velocity vector of the incoming particles would rotate over a complete 2π-steradian solid angle for a complete orbit. A "nude" type ion source was designed which employed the orthogonal orientation of the velocity vector and gage axis to minimize the surface interactions of the ambient particles before measurement.

The need to focus multi-directional particles with energies up to 12 ev with high sensitivity resulted in both an unusual ion source and a special type of analyzer. The characteristics of the equipment were: a large solid angle of acceptance, a short ion path length, and a large tolerance to forward energy variations.

The preceding characteristics were achieved by the use of a double focusing instrument where the ions were dispersed along a focal plane, and the electric and magnetic fields were kept constant.

Six ion collectors were positioned along the focal plane to detect the preselected masses. These were sampled in sequence by a switching arrangement in the amplifier. Provisions were also made to sample periodically the total ion current in the analyzer region.

The arrangement of the various components is shown in Figure 3.39. There are separate detectors for mass numbers 4 (He^+), 14 (N^+), 16 (O^+), 18 $(H_2O)^+$, 28 (N_2^+), and 32 (O_2^+). A two amplifier detector system is employed to fulfill the requirements of large dynamic range, low current detection and maximum accuracy.

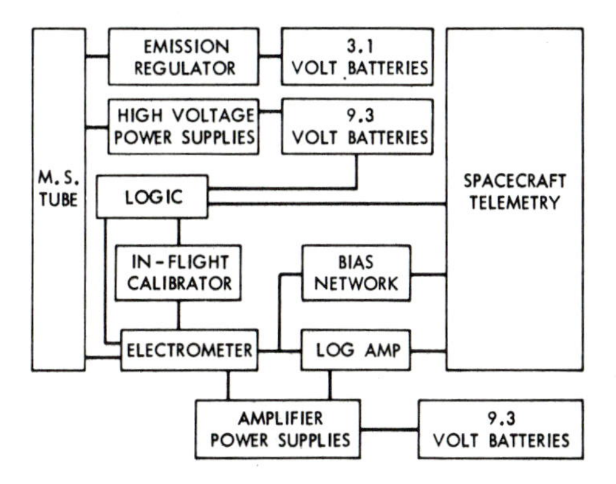

Figure 3.40 Block diagram of the mass spectrometer system

A block diagram of the mass spectrometer is shown in Figure 3.40. The low currents are detected with a sensitive electrometer amplifier, and the output is fed to a logarithmic amplifier matched to the output telemetry.

Results

A spin stabilized Explorer was placed into orbit in 1963 at a 58° inclination, 250 km perigee, and a 900 km apogee. The satellite, carrying two mass spectrometers, was a stainless steel sphere, 35″ in diameter. The two mass spectrometers described above were located at opposite ends of the spin axis. Although the spectrometers functioned for only part of the total satellite lifetime, it was possible to demonstrate that direct measurements of atmospheric composition are possible from a high altitude Earth orbiting satellite. Successful measurements were made of major components of the neutral atmosphere over a 2 month period. A large variability in the number densities at a given altitude was observed. This indicated a strong sensitivity to changes in energy input to the atmosphere, particularly changes in magnetic activity.

Lunar Atmosphere Mass Spectrometer (Hoffman, 1969)

J. H. Hoffman of the Southwest Center for Advanced Studies has proposed a lunar atmosphere mass spectrometer as part of a payload for a lunar orbiting vehicle such as a separate satellite or the Apollo service module. The purpose for the experiment is to investigate a number of gases from orbital altitudes. The particular gases are the noble gases, carbon dioxide, carbon, hydrogen sulfide, ammonia, sulfur dioxide and, water vapor, conceivably released by lunar volcanism from rocks and magma. It is suggested by Hoffman that these measurements would yield important information on the processes listed in Table 3.4

Table 3.4. *Lunar orbiter mass spectrometer objectives*

1. Volcanism and release of gases from surface.
   A. Physical processes causing release.
   B. Chemistry of moon.
2. LM Rocket gas escape rate.
   A. Diffusion rates of gas cloud.
   B. Outgassing rate of surface.
3. Global behavior of gases.
   A. Concentration of atmospheric gases on night side.
   B. Frozen gases release at dawn terminator.
4. Solar wind accretion.
   A. Neon distribution.
   B. Detection of hydrogen and helium.

Instrumentation

Two types of mass spectrometers have been developed to meet the differing constraints of weight, power, and telemetry bandwidths. Both instruments are basically similar, employing magnetic sector-field mass spectrometers with a Nier-type thermionic electron bombardment ion source. A diagrammatic sketch of both analyzers is shown in Figure 3.41. The less elaborate instrument has two emerging ion beams from the magnet, whereas the larger instrument has three emergent ion beams. The dual beam instrument has two collector systems adjusted to scan, simultaneously, mass ranges from 1 to 4 amu and 12 to 66 amu with the possibility for adjustment to include krypton up to mass 86. The triple beam instrument has 3 collector systems for the simultaneous scanning of the mass ranges 1 to 4 amu, 12 to 48 amu, and 40 to 160 amu. The

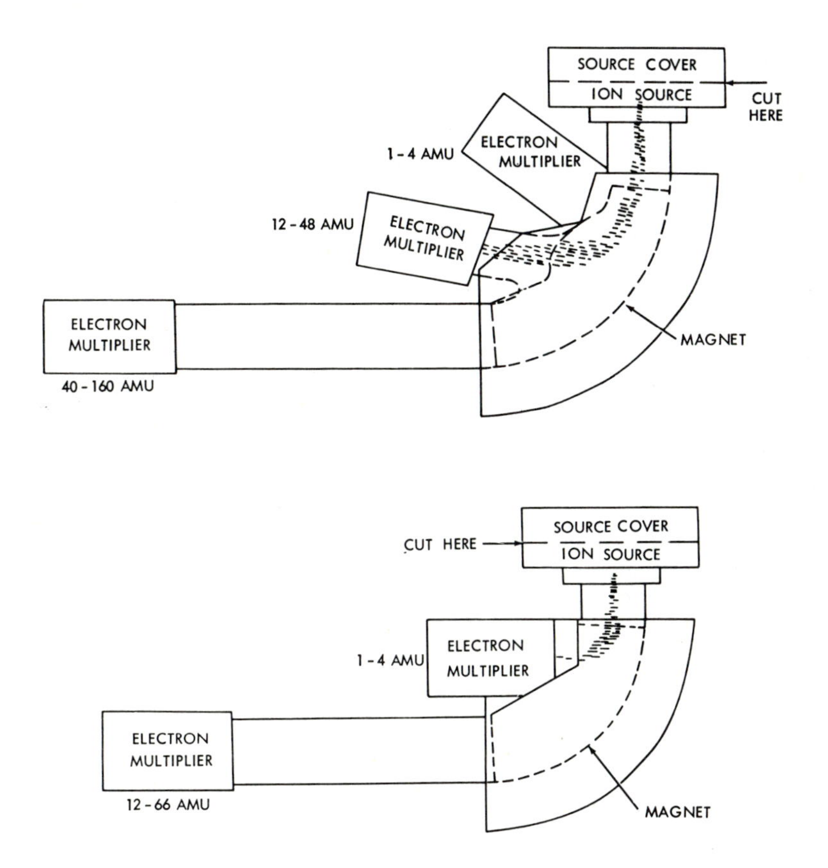

Figure 3.41 Analyzer diagrams

various ion beams traverse the magnetic fields with different radii of curvature.

The operation of the instrument depends on gases entering the ion source chamber which is maintained at a high positive potential. Ions are formed in the source chamber by electron bombardment. The gas ions produced are accelerated through a pair of exit slits, forming a collimated beam which enters the magnetic analyzer at ground potential. The ion beam passes through a uniform magnetic field which separates the ions into three different trajectories (for the three beam system). Those ions meeting the focus conditions pass through a collector slit and are counted. Three separate counter systems are used, and the spectrum is scanned by varying the ion accelerating voltage. The advantage of the three beam system is that a much smaller range of accelerating potentials is required.

The ion source is a Nier-type thermionic electron bombardment source. Ions are formed by a collimated electron beam from a hot filament at 75 ev energy. Either of two filaments are selectable on command, and the emission current is monitored as a housekeeping function.

The ion accelerating high voltage supply is operable in two modes: analog and digital. The analog mode provides an RC decay voltage with periods ranging from 30 to 60 sec. In the digital mode, a step function is generated which approximates an exponential decay of the voltage.

A clock, synchronized to the frame rate of the spacecraft's telemetry system, is used to operate an 11 bit counter. The counter's output is converted to analog information which, in turn, is used to control a high voltage power supply; its output is a series of voltage steps which approximates an exponential wave form from 1600 to 400 V in 1958 voltage steps. For the three beam instrument, this results in scans of 1 to 4 amu, 12 to 48 amu, and 40 to 160 amu, the planned sweep time is about 196 seconds. Background counting rates will be obtained between successive sweeps.

The collector system used depends on the manner of operation; several types are possible. In the first system, an electron multiplier-log electrometer amplifier is used to collect the beam. The spectrum is swept in periods of 30 to 60 seconds, and the collected ion currents are converted from analog to an eight-bit digital word which is fed into a "Peaks" circuit. The Peaks circuit stores both the amplitude and position of each peak for subsequent telemetry. An amplifier-"Peak" system is used for each collector.

A second collector system uses an electron multiplier behind each collector slit as a preamplifier. In this mode, single ions arriving at the collector slit are counted. This method has a dynamic range of about six orders of magnitude. The counting system is synchronized to the clock

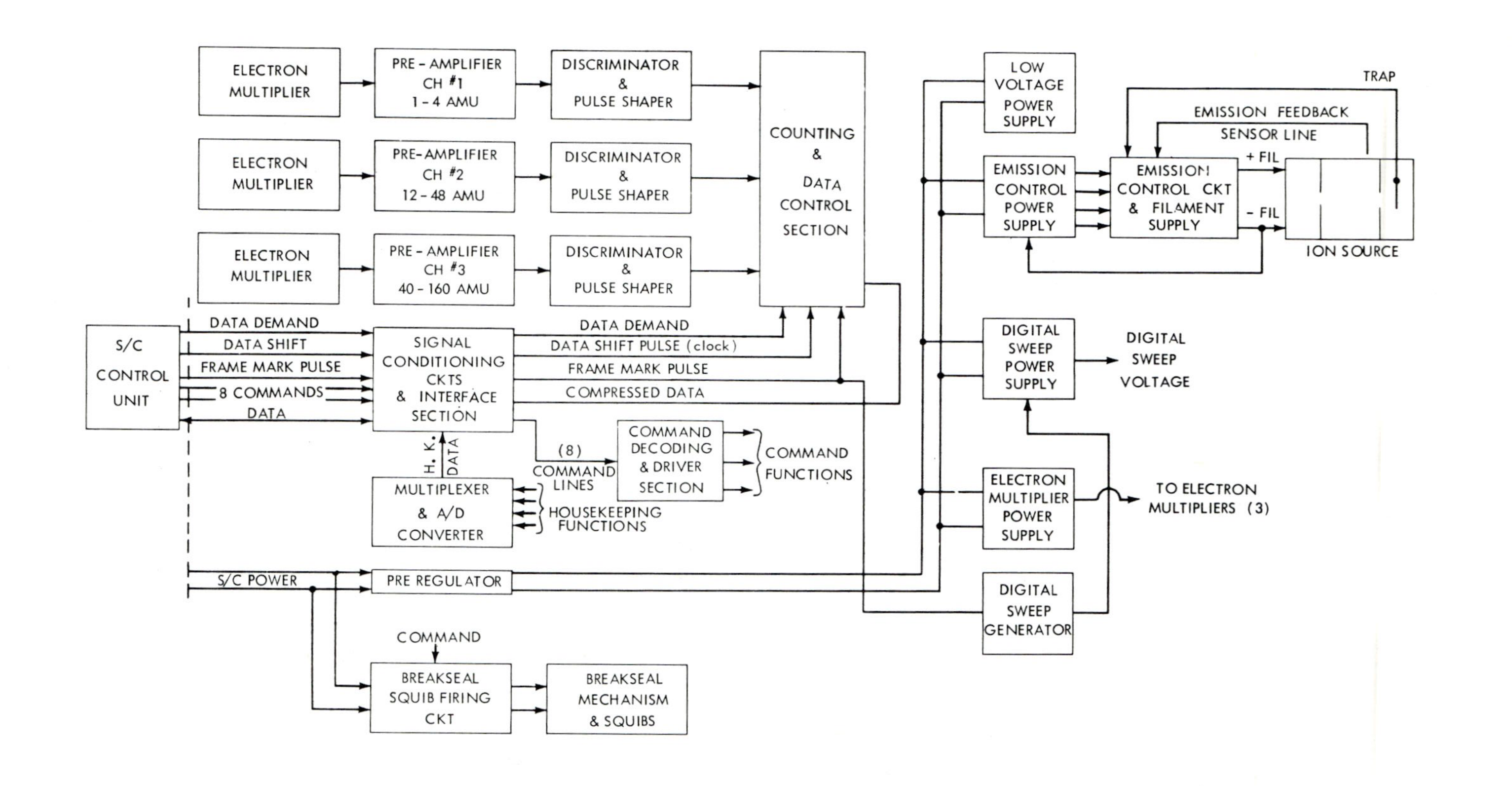

Figure 3.42 Lunar mass spectrometer functional block diagram

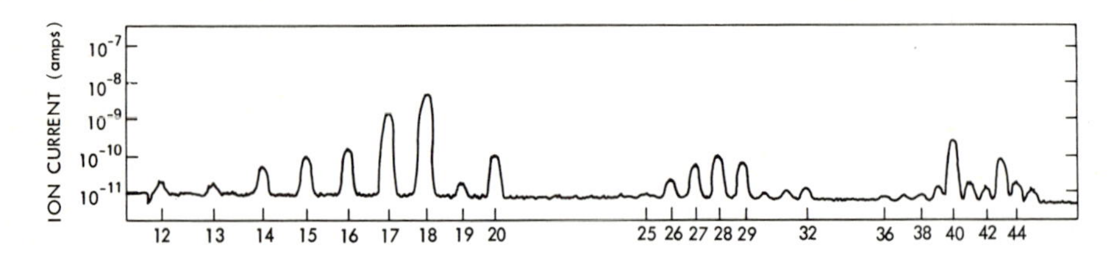

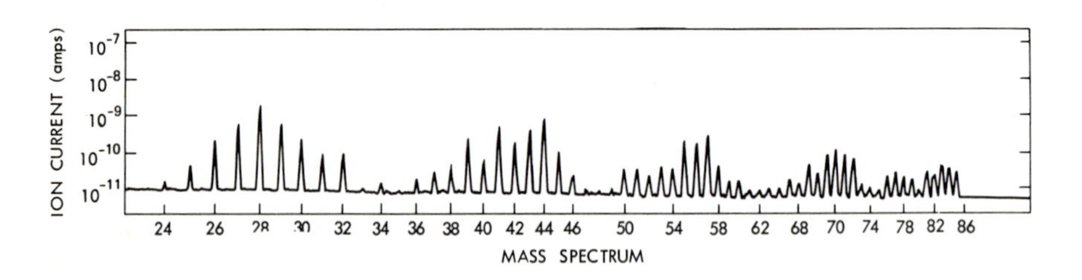

Figure 3.43 Spectra obtained with a laboratory model of the triple beam instrument

Table 3.5 *Mass spectrometer specifications mechanical*

Property	Double collector	Triple collector
1. Mass range (amu)	1–5, 12–66	1–4, 12–48, 40–160
2. Resolution	5, 60	4, 45, 130
3. Sensitivity		
A. Peaks	10^{-12} torr	10^{-12} torr
B. Step	10^{-13} torr	10^{-13} torr
C. Lock	3×10^{-14} torr	3×10^{-14} torr
4. Sweep time		
A. Peaks	40 sec	30 sec
B. Step	180 sec	200 sec
C. Lock	100 sec	120 sec
5. Weight (lbs)		
Magnet	3	4
Total	10–13	14–17
6. Size	400 cu. in. 1 dim. 10 in.	600 cu. in. 1 dim. 12 in.

which drives the high voltage sweep circuit, and the number of counts accumulated during a voltage step is obtained by each counting system.

Still another option offered in the Hoffman device is called the "Lock" mode. Again a stepping voltage scan is used, but with shorter steps. As soon as a peak is detected, the sweep locks onto the peak for several tenths of a second while the number of ions passing through the collector system are counted and telemetered. Figure 3.42 shows a functional block diagram of the electronics, and Table 3.5 lists the mechanical and electrical characteristics of the equipment.

Figure 3.43 shows two examples of spectra obtained with a laboratory model of the triple beam instrument. The upper spectrum is an intermediate range spectrum with the mass collector operated in an analog mode. The spectrum is of a mixture of argon and a background of hydrocarbons. The lower spectrum covers the mass range from 24 to 86 amu and is also a mixture of hydrocarbon peaks.

Mass Spectrometer for the Analysis of Solids (Hertzog, 1965, Kreisman, 1969)

Apart from the mass spectrometer developments described above, the following device has been designed as a possible lunar probe for the direct analysis of solids, either in situ or as a small laboratory device to be used at some future lunar station. Although the present equipment does not permit precise enough analysis for the determination of isotope ratios, it does offer potential for rapid chemical analysis with high sensitivity for major and trace elements—particularly those of low atomic number. The miniaturized solids mass spectrometer, for which a breadboard version now exists, uses a primary beam of positive ions to bombard the sample surface under analysis. Many of the atoms and molecules of the sample are sputtered from the target material, some in the form of ions. These are sorted by a small, double focusing, magnetic deflection mass spectrometer.

A block diagram of the mass spectrometer is shown in Figure 3.44. There are three major basic components: an ion source, an analyzer section, and an ion detector. Because of limited electrical power, it was necessary to exclude the use of an electromagnet and to use a fixed magnet, instead, in the analyzer. The use of a permanent magnet made it necessary to divide the desired mass range of 1 to 100 amu into two separate mass ranges: a low range of 1 to 12 amu, and a high mass range from 10 to 100 amu. In the breadboard instrument, the separation of the two beams was accomplished by collecting the lighter low mass ions after a short trajectory through only a small portion of the magnetic field. Both mass

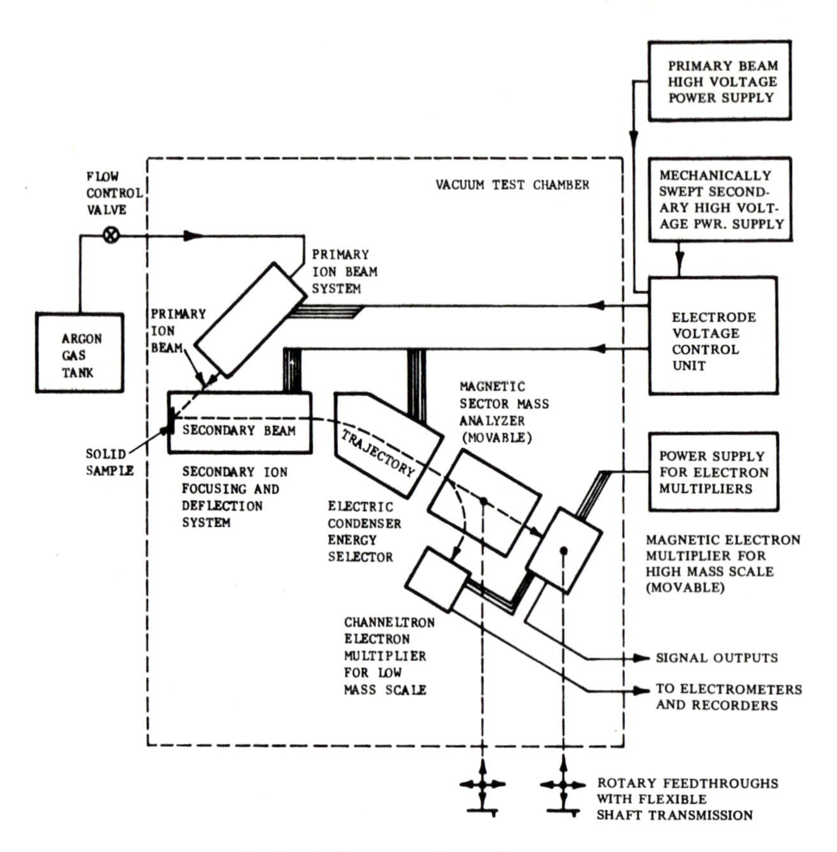

Figure 3.44 Schematic block diagram of lunar lander mass spectrometer

ranges are covered simultaneously by a single sweep of the secondary ion beam accelerating voltage; both beams are measured by means of an electron multiplier properly placed to intercept both the high and low mass beam.

The sputter ion source, which consists of several elements, is shown in Figure 3.45. Primary beam ions of argon are formed in a Penning-type cold cathode discharge source. The positive ions are then extracted from the discharge region and focused by an einzel or unipotential lens to form a small spot at the sample. The various elements are: (2) Penning source anode, (1) and (4) the cathode, (3) a two-piece sleeve or ring magnet to supply the magnetic field for the Penning source, (14) a tube for the primary beam gas, (5) the extraction electrode, (6), (7) and (8) the three elements of the einzel lens, and (9) the sample holder.

In practice, the Penning source cathodes are operated at the full primary beam positive voltage. The Penning source anode is operated

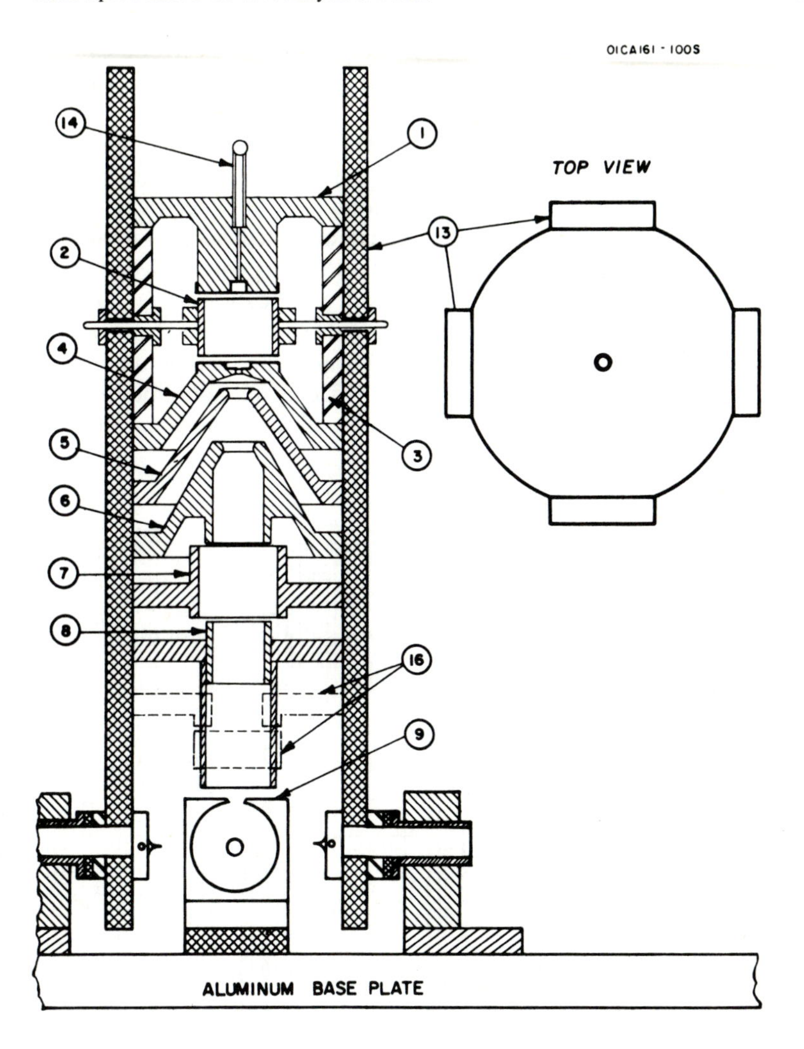

Figure 3.45 Schematic of lunar lander mass spectrometer primary ion beam system

at about 500 to 1000 V. The anode to cathode potential can be smoothly varied from 0 to 100 V by means of a variac. The extraction electrode is operated at a few thousand volts negative with respect to the source cathode. The two outer electrodes of the einzel lens are maintained at ground potential, while the center electrode is kept at a positive potential in order to function as a "decel-accel" lens. This type of lens provides strong focusing over a short distance. The extraction electrode can be varied in potential to change the ion beam intensity without seriously affecting the beam focus.

Observations on this Penning discharge showed that stable performance could be expected for magnetic fields of under 800 gauss, and currents between 6 and 30 ma. It was also found that the spectra from this ion source compared favorably with those obtained using a standard duoplasmatron source. The advantages of a Penning source over that of a duoplasmatron source for possible space application were: no hot filament was required, the construction was simpler, and the power consumption was less. The application to a non-conducting material is shown in Figure 3.46 for aluminum oxide. Once again, note both atomic and molecular species. The silicon line represents an obvious impurity in the aluminum oxide.

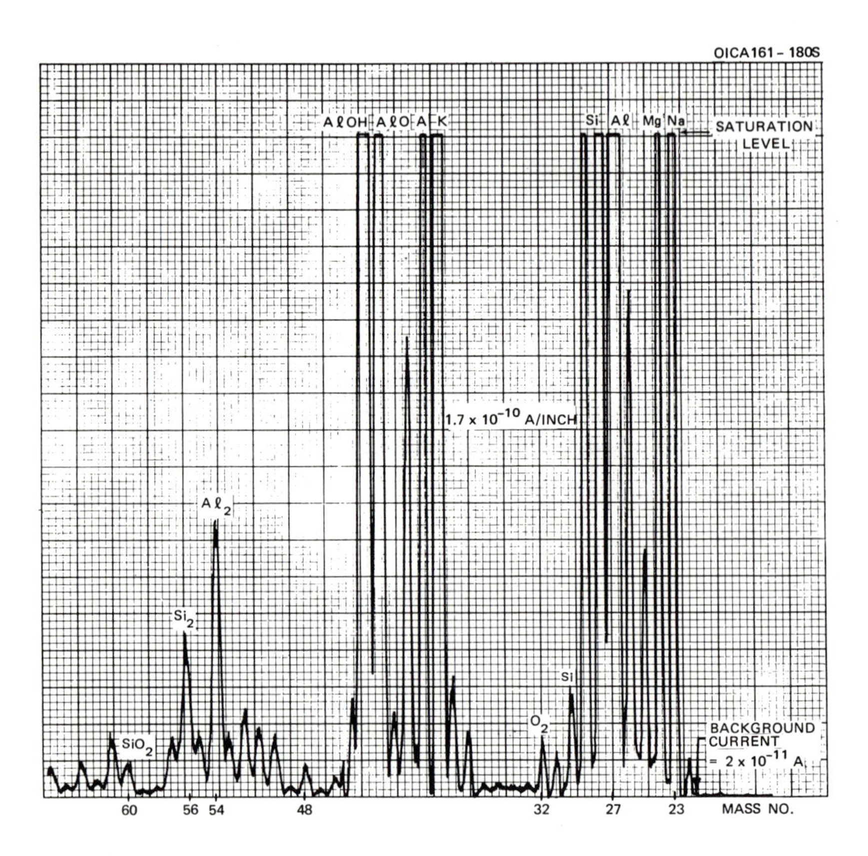

Figure 3.46 Mass spectrum of pure aluminum oxide

In summary, the above device represents a promising beginning to the design of a compact mass spectrometer for possible use in advanced lunar missions. The resolution of the breadboard instrument is about 75, but resolutions of about 100 can be achieved with narrower slits. The sensitivity for a low atomic number element such as silicon is about 10 ppm or better. There are still problem areas for future exploration. These are the chargeup of insulating materials, and poor resolution and sensitivity for the very low atomic masses. The latter problem is an instrumental one related to the inhomogeneity of the magnetic field through which the low mass ions pass, and the interference of the magnetic fringe field with the electron multiplier. The former problem of sample chargeup can seriously affect the application to soil materials. A possible solution may lie in an appropriately designed sample holder to carry off surface charge. The various components of the secondary beam system are shown and identified in Figure 3.47. This drawing is not to scale and is meant only to identify the various components.

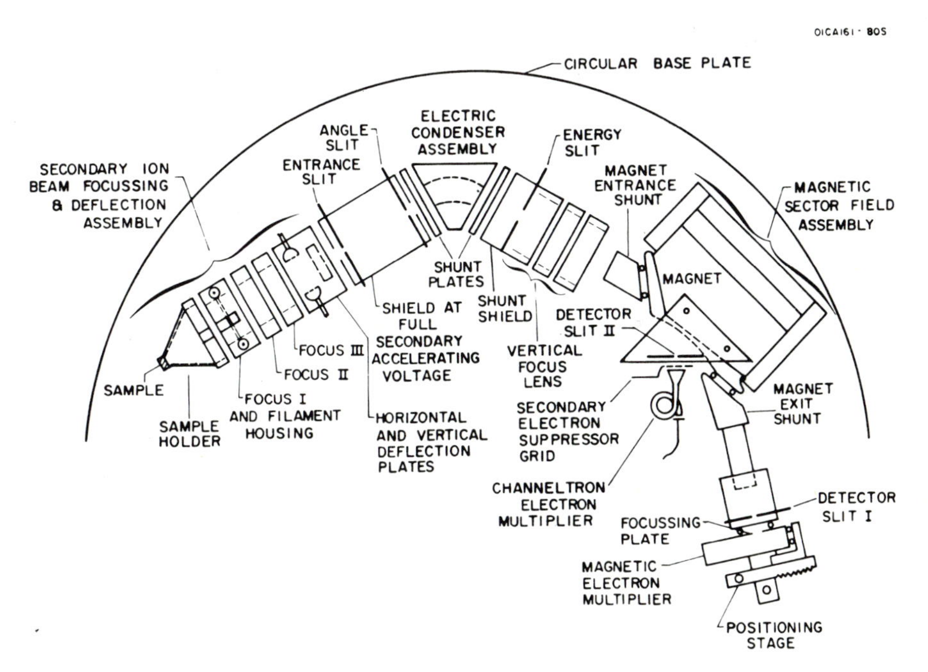

Figure 3.47 Components of the secondary beam system

High energy primary beam ions striking the sample surface cause atoms and molecules of the surface to be sputtered in all directions. Because the primary ion beam energy is relatively high, there is a signifi-

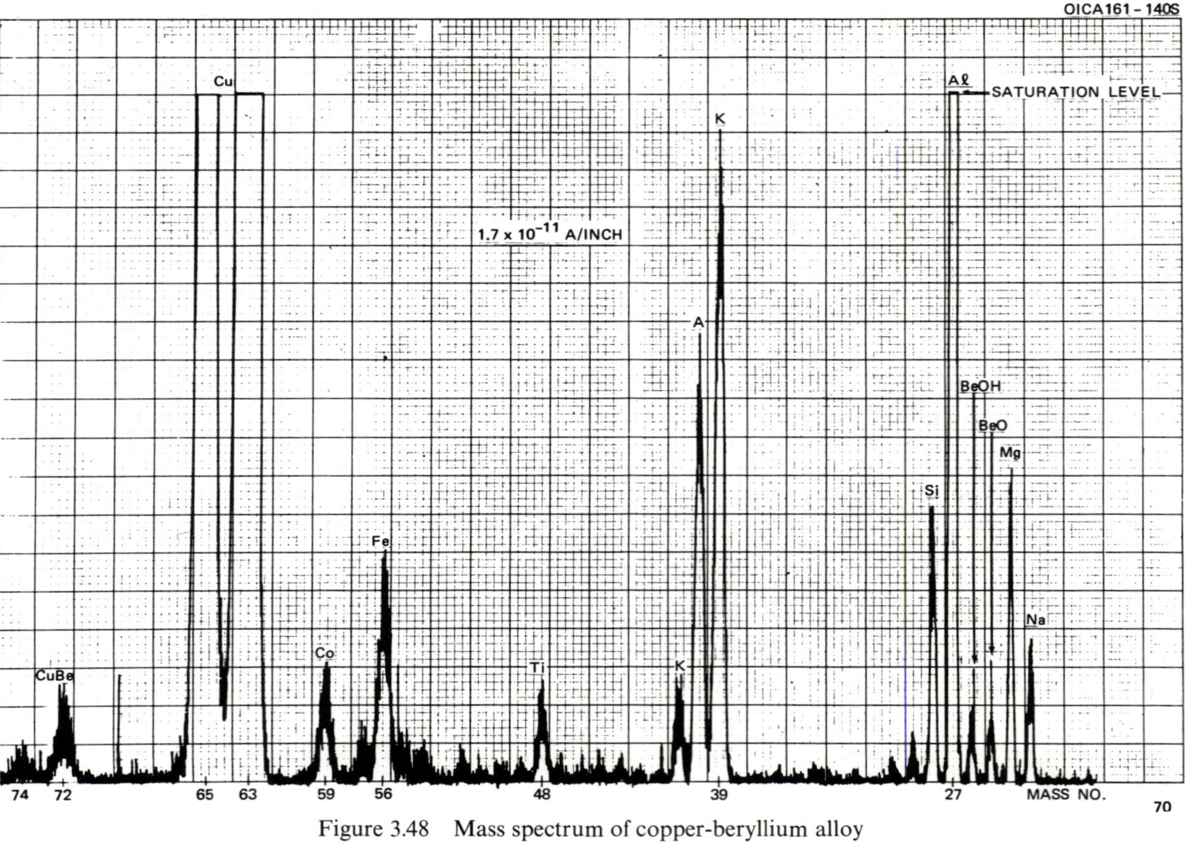

Figure 3.48 Mass spectrum of copper-beryllium alloy

cant number of positive ions formed. These "secondary ions" from the sample are emitted in all directions, with some characteristic distribution of velocities or energy. The secondary ion beam system, containing an electric and magnetic analyzer, is designed to extract, focus, and mass analyze the seondary ions. For a discussion of the sputtering process, the reader is referred to the reports of Herzog (1965), and Kreisman (1969).

Results

Some preliminary results showing the potential of this instrument are shown in Figure 3.48. The spectrum for the copper-beryllium alloy covers the atomic mass range from approximately 10 to 75. Adjacent mass numbers can easily be resolved up to about mass 70. It can also be seen that the spectrum is made up of both atomic and molecular species.

Gas Chromatography

The application of gas chromatography to the analyses of volatile constituents on the lunar surface was suggested as far back as 1961 (Oyama et al., 1961); the technique was then studied at JPL and Aerojet-General Corp. A considerable amount of effort has gone into the design of the equipment, and an elaborate device is now under development for the Viking mission to Mars scheduled for 1975.

Gas chromatography is a method for separating materials in the vapor state by a process of sorption-desorption from solids. A flow

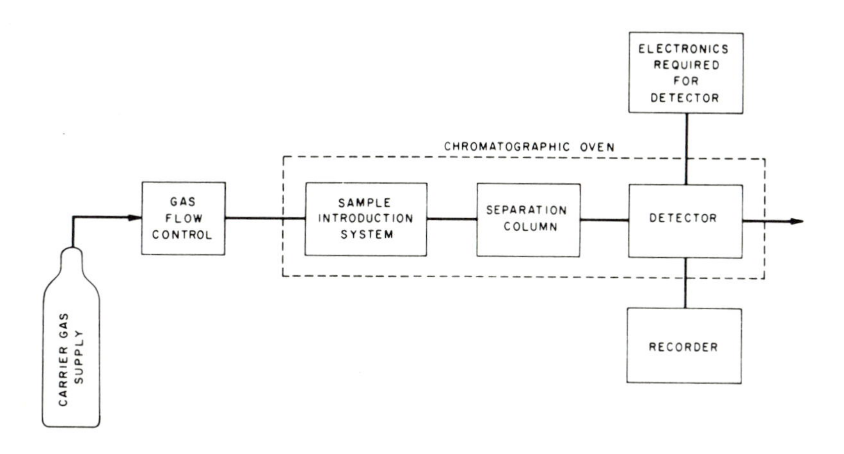

Figure 3.49 Schematic diagram of a gas chromatograph

diagram of a typical gas-chromatography apparatus is shown in Figure 3.49. Mixtures of gases or vaporized solids are introduced into a column packed with an appropriate adsorbent. An inert carrier gas such as helium then carries the sample vapor through the adsorption column. As the mixture passes through the column, various components are retained and then desorbed according to their physical characteristics. At some distance from the initial point of introduction the components are separated. The desorbed materials may be trapped as they flow from the column, or be detected by various types of detectors or sensing systems. The gas chromatograph can also be coupled to a mass spectrometer in order to determine the emerging species.

The detecting systems may be used to quantitatively determine the various components in a mixture. The nature of the materials determines the retention time. This, in turn, can be used to identify the specific components.

Gas chromatography is potentially a powerful technique for the detection of volatile constituents on the lunar surface, the identification of compounds of possible biogenic origin, and the analysis of planetary atmospheres.

Lunar Gas Chromatograph

The following device was developed for the purpose of determining the presence of organic constituents on the lunar surface. This work was done as a JPL effort, aided by S. R. Lipsky of Yale University and J. E. Lovelock of Baylor University. Figure 3.50 shows a schematic drawing of the lunar gas chromatograph.

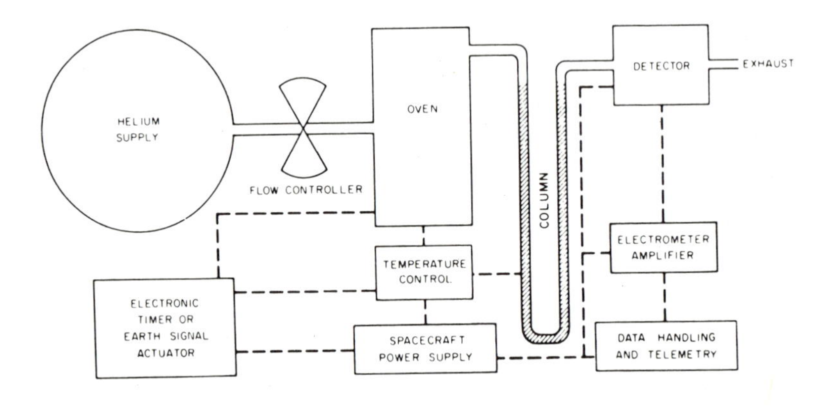

Figure 3.50 Schematic drawing of the lunar gas chromatograph

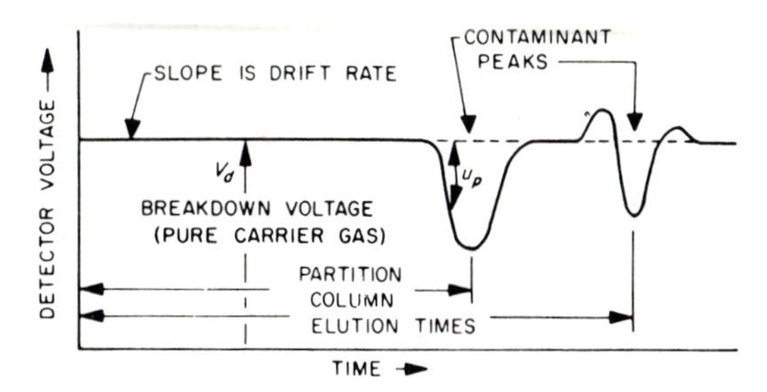

Figure 3.51 Typical glow-discharge detector signal

Table 3.6. *Performance requirements for the lunar gas chromatograph*

Components to be resolved	
Hydrogen	Propionaldehyde
Oxygen	Formic acid
Nitrogen	Acetic acid
Carbon monoxide	Butyric acid
Carbon dioxide	Benzene
Methane	Toluene
Ethane	Acetone
Propane	Acetonitrile
Butane	Acetylene
Methanol	Acrolein
Ethanol	Hydrogen cyanide
Propanol	Hydrogen sulfide
Formaldehyde	Ammonia
Acetaldehyde	Water

Performance requirements	
Maximum retention time, min	10
Minimum detectable quantity in oven, mole	3×10^{-10}
Minimum dynamic range of detection	10,000 times minimum detectable quantity
Oven temperature control, °C	± 10
Oven maximum heating time, min	4
Detector and Signal processing:	
Output, V	0 to 5
Maximum noise level (peak to peak), mV	100
Minimum detectable signal twice the maximum noise level accuracy, %	± 1
Oven seal maximum leakage (helium), cm^3/sec	$\pm 10^{-6}$
Column temperature control, °C	± 0.25
Valve maximum leakage (helium), cm^3/sec	10^{-4}
Retention time reproducibility, %	1
Pressure regulation, %	1
Calibration sample reproducibility, %	4

The instrument as shown, constructed by Beckman Instruments, Inc., weighed 14 pounds and occupied a volume of 640 cb. in. The power required for a 100 min analysis was 24 Wh.

This equipment was planned as part of a Surveyor payload, to operate in automated mode after the Surveyor's arrival on the lunar surface. Conceptually, samples were to be picked up by a sampling device, and delivered through a funnel to the gas chromatography oven. The oven was programmed for three temperatures, 150°, 325°, and 500° (depending on an uplink signal from the earth). The carrier gas was helium, and the detector, was a glow discharge type. The various components were determined from the breakdown voltages of the helium carrier gas which was to be telemetered. A typical glow discharge detector signal is shown in Figure 3.51.

Some idea of the difficult performance requirements are shown in Table 3.6.

Planetary Gas Chromatographs

A schematic diagram of a laboratory model of a gas chromatograph for planetary studies is shown in Figure 3.52. This apparatus was reported on by Chaudet (1962), and was developed by Melpar, Inc. for JPL. The instrument is designed to quantitatively determine the gases listed in Table 3.7.

Table 3.7.

Components to be determined in planetary gas chromatograph		
Hydrogen	Krypton	Ethane
Oxygen	Methane	Carbon dioxide
Argon	Nitrogen	Carbon monoxide
Water vapor	Ammonia	Nitrous oxide

Table 3.8. *Functions of the various columns in planetary gas chromatograph*

Column	Function
1	Air, NH_3, H_2O
2A	Delay passage of air to detector 2 until NH_3 is eluted from column 1
2B	Kr, Xe, C_2H_6, CO_2
3	H_2—Ne, A—O_2, N_2, CH_4, CO, Xe
4	To delay elution into detector 4 to prevent interference with detector 3

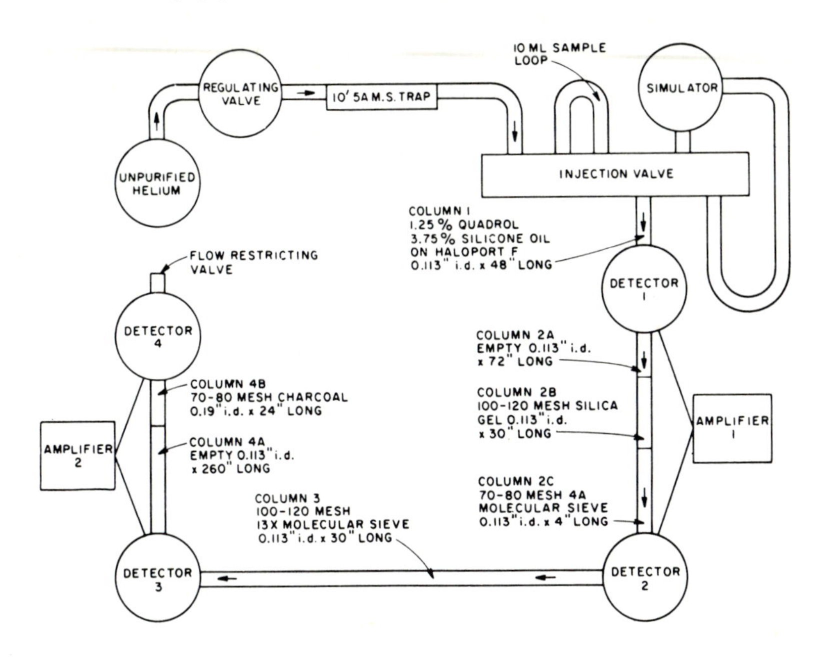

Figure 3.52 Overall schematic of planetary gas chromatograph

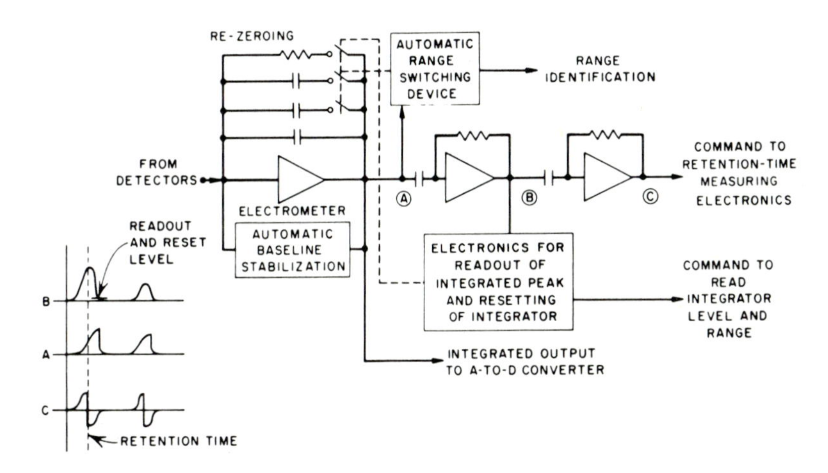

Figure 3.53 Partial block diagram of peak-integrating electronics

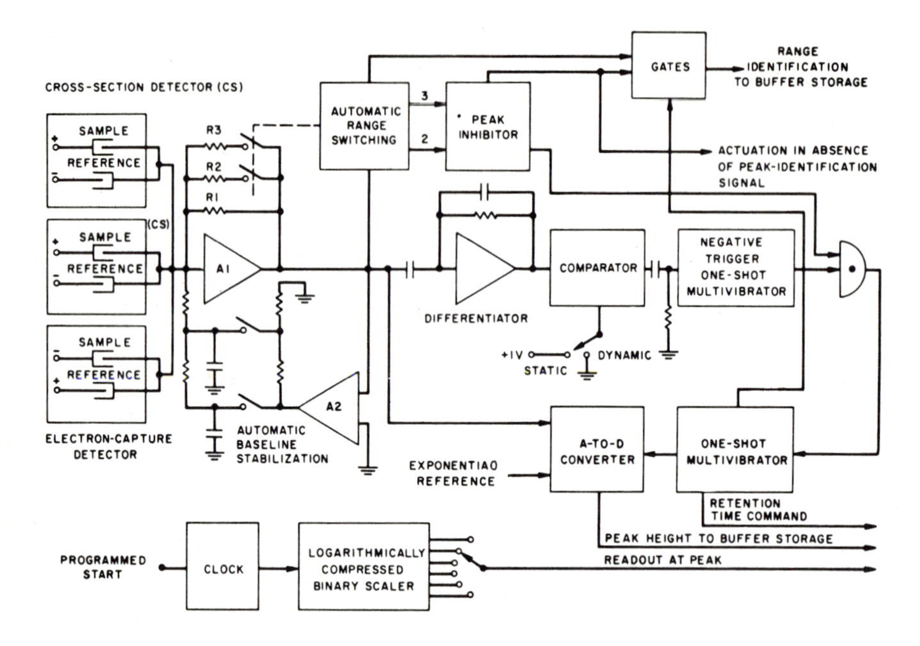

Figure 3.54 Peak-height measurement system

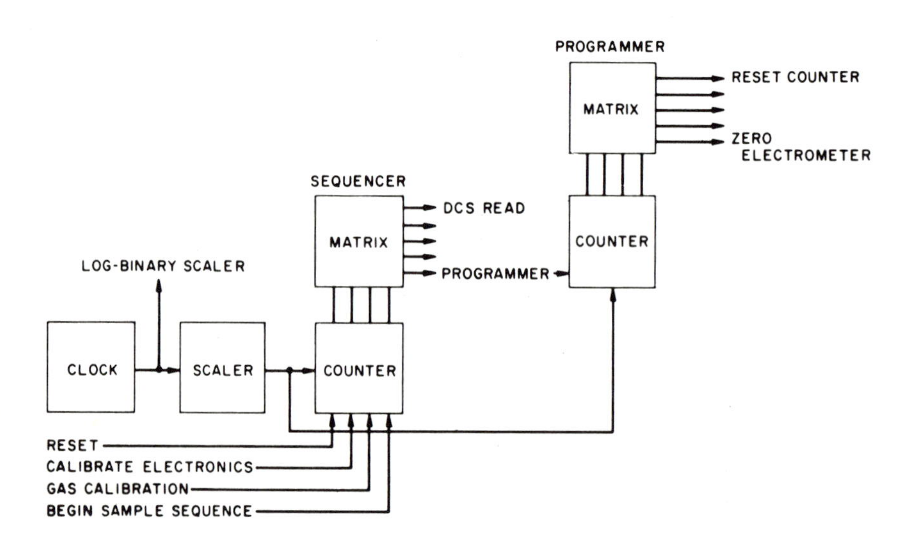

Figure 3.55 Gas chromatograph sequences and programmer

The chromatograph design involves a complex arrangement of series connected micro-cross-section detectors and separation columns. Unusual features of the absorption column system are the use of charcoal (shown as 4 B) to absorb oxygen irreversibly in order to separate oxygen from argon, and the use of a short moelcular sieve (Col. 2 C) to separate nitrous oxide and carbon dioxide. A summary of the use of the various columns is given in Table 3.8.

A unique feature of the JPL/Melpar gas chromatography is the use of a chemical-cartridge heating system in the oven assembly for pyrolyzing the sample. The cartridge is quite small and uses a Zr—$BaCrO_4$ primer on the bridge wire, an Al—Fe_2O_3 booster, and a delay line charge. The oven temperatures can be elevated from 200 to 310 °K for about an hour by the sequential firing of the 12 cartridges surrounding the oven.

A working model of a planetary gas chromatograph for a Mars mission was undertaken as a joint effort of Lipsky (Yale University) and JPL (1963). The chromatograph was to be operated during a descent to the Martian surface, and was to be capable of performing a complete gas analysis during a 60 sec descent.

The electronics for performing quantitative measurements are shown in Figure 3.53 (Josias, 1964). Included are: an electrometer with a five-decade range for the linear reproduction of all detector peaks, an electronic integrator with a minimum range of 6.5 decades. Figure 3.54 is a block diagram of the peak height measurement system, and Figure 3.55 shows the use of the electrometer amplifier as an integrator for measurement of peak areas as opposed to peak heights. Figure 3.55 also shows the method for sequencing the operation of the gas chromatograph for completely automatic control.

Combined Gas Chromatography—Mass Spectrometry

The possibility of using a combination gas chromatograph and mass spectrometer to improve the analysis of organic components has received considerable attention. Bentley (1965) has pointed out that the identication of the components of a mixture, where the components have similar mass spectra, can be done much more effectively if some knowledge of gas retention time by a gas chromatographic column is added.

Now under development for the Viking 75 mission, is a system intended to perform remote gas analysis on the Martian surface. Because the device is still in a developmental stage, it is not possible to describe it in detail. The instrument will, however, use pyrolysis to evolve gases from samples delivered to the various electrically heated pyrolysis ovens. The ultimate design will quite probably combine the JPL gas chromatograph (see above) with a small mass spectrometer (either a small magnetic sector or quadrupole).

The requirements for such a device are certainly very rigorous. It must function reliably in a highly programmed mode, yet transmit unambiguous information. Further, it will have to survive stringent sterilization procedures, and be rugged enough to survive a substantial impact stress on landing.

References

Adler, I., Trombka, J. I.: An integrated experiment for compositional analysis from an orbiter. Advanced Space Experiments, Vol. 25. American Astronautical Society 1969.

— — Rock analysis by alpha excitation of X-rays, a possible lunar probe, presented at 2nd Symposium on Low Energy X and Gamma Sources and Applications, Austin, Texas (March, 1967). Proceedings published by U.S.A.E.C. as O.R.N.L.-11C-5, 1968.

Aronson, J. A., Emslie, A. G., Allen, R. V., McLinden, H. G.: Far infrared spectra of silicate minerals for use in remote sensing of lunar and planetary surfaces. NASA Contract Report, NAS-8-20122, 1966.

Bentley, K. E.: Detection of Life-Related Compounds on Planetary Surfaces by Gas Chromatography-Mass Spectroscopy Techniques. JPL, TR 32—713 Aug. 30, 1965.

Caldwell, R. L., Mills, W. R., jr., Allen, L. S., Bell, P. R., Heath, R. L.: Combination neutron experiment for remote analysis. Science **152/3721,** 457 (1966).

Cameron, J. F., Rhodes, R. J.: X-ray spectrometry and radioactive sources. Nucleonics **19/6,** 57 (1961).

Chadwick, J.: Philosophical Magazine **24,** 594 (1912).

Chaudet, J. H.: Gas chromatographic instrumentation for gas analysis of Martian atmosphere. Melpar Inc. Final Report, NASA-CR 59773, 1962.

Das Gupta, R., Schnopper, H. W., Metzger, A. E., Shields, R. A.: A combined focusing X-ray diffractometer and non-dispersive X-ray spectrometer for lunar and planetary analysis. Advances in X-ray Analysis **9,** 221, Plenum Press, N. Y., 1966.

— Pan, N.: A new experimental technique for X-ray diffraction studies. J. Sci. Ind. Res. (India), No. 178, 131, (1958).

Friedman, H.: Advan. Spectr. Ed. by Thomson, Wiley & Sons, 1964.

Goetz, A. F. H., Westphal, J. A.: A method for obtaining differential 8—13 micron spectra of the moon and other extended objects. Appl. Opt. **6,** 1981 (1967).

— Differential infrared lunar emission spectroscopy. J. Geophys. Res., (1455), 1968.

Gorenstein, P., Mickiewicz, S.: Reduction of cosmic rays observed on OSO-1. University of California, San Diego, Special Publication 68-1, 1967.

Greenwood, R. C., Reed, J. H., Kolar, R. P., Terrel, C. W.: Trans. Am. Nuclear Soc. **6,** 181 (1963).

Herzog, R. F. K.: Solids mass spectrometer. GCA Corp., Final Report, NASA Contract NA Sw-839, 1965.

Hoffman, J. H.: Private communication, 1969.

Hovis, W. A., Callahan, W. R.: Infrared reflectance of igneous rocks, tuffs and red sandstones from 0.5 to 22 microns. J. Opt. Soc. Am. **56,** 639 (1966).

— Lowman, P. D.: Private communication, 1968.

Imamura, H., Vehida, K., Tominaya, H.: Fluorescent X-ray analyzer with radioactive sources for mixing control of cement raw materials. Radioisotopes 11(4), July 1965.

Josias, C., Bowman, L., and Mertz, H.: A Gas Chromatography for the Analysis of the Martian Atmosphere. JPL, Space Programs Summary 37—24, vol. VI, Dec. 31, 1963, p. 213.

Kartunnen, J. O.: A portable fluorescent X-ray instrument utilizing radioisotope sources. Anal. Chem. **36,** 1277 (1964).

Kendall, B. R. F., Herzog, L. F., Wyllie, P. J., Emonds, D. S., Bauer, C.: Analysis of the lunar surface and atmosphere by mass spectroscopy. NASA Contract, NAS 8-11119, 1964.

Kreisman, W. S.: Lunar Lander Mass Spectrometer, Phase 11. GCA Corp. Final Report, Contract NAS 5-9393, 1969.

Lee, H. B., Bert, L. E., Wainerdi, R. E.: The use of neutron activation analysis to determine the elemental composition on the moon's surface. Trans. Am. Nucl. Soc., 5 (2), 1962.

Lyon, R. J. P.: Evaluation of infrared spectrophotometry for compositional analysis of lunar and planetary soils, NASA Contractor Report CR-100, 1964.

— Analysis of rocks by spectral infrared emission (8 to 25 microns), Econ. Geol. **60,** 715 (1965).

Lipsky, S. R., Bentley, K., Bowman, L., Josia, C., Mertz, H., Wilhite, F.: Mariner B 1966 mission, analysis of the atmosphere of Mars by gas chromatography instrumentation, Proposal to Jet Propulsion Laboratory by S. R. Lipsky, Nov. 15, 1963.

Mack, M., Parrish, W.: Seeman-Bohlin X-ray diffractometry: instrumentation, 23rd Pittsburgh Diffraction Conference, Mellon Institute Paper B 3, Nov. 3, 1965.

Merzbacher, E., Lewis, H. W.: X-ray production by heavy charged particles, Handbuch der Physik, 34 (166), Berlin-Göttingen-Heidelberg: Springer 1958.

Metzger, A. E.: Some calculations bearing on the use of neutron activation for remote compositional analysis. Jet Propulsion Laboratory Technical Report No. 32-266, Jet Propulsion Laboratory, Pasadena, California, 1962.

— An X-ray spectrograph for lunar surface analysis. NASA Contractor Report-609340, 1964.

Mills, W. R., Givens, W. W., Caldwell, R. L.: A combined pulsed neutron experiment for elemental analysis of lunar and planetary surfaces. Mobil Research and Development Corporation, Dallas, Texas, 1969.

Monaghan, R., Younous, A. H., Bergan, R. A., Hopkinson, E. C.: I.E.E.E. Inst. Electr., Electron Eng. Trans. Nuclear Science, **10/1,** 183 (1963).

Nash, D. B.: Lunar and planetary X-ray diffraction program. Jet Propulsion Laboratory Technical Memorandum No. **33/218,** 51 (1965).

Oyama, V. I., Vango, S. P., Wilson, E. M.: Applications of gas chromatography to the analyses of organics, water and adsorbed gases in the lunar crust. Jet Propulsion Laboratory Technical Report **32/107** (1961).

Parrish, W.: Lunar X-ray diffractometer. Final Report, Jet Propulsion Laboratory Contract No-950158, July 1964.

Reber, C. A., Nicolet, M.: Investigation of the major constituents of the April–May 1963 heterosphere by the Explorer XVII satellite. Planetary, Space Sci., 617, 1965.

Robert, A.: Contributions to the analysis of light elements using X-ray fluorescence excited by radioelements. Commissariat a L'Energie Atomique, Rapport CEA-R 2539, 1964.

Robert, A., Martinelli, P.: Method of radioactive analysis of heavy elements by virtue of X-ray fluorescence. Paper No. SM 55/76 at IAEA Meeting, Salzburg, Austria, October 1964.

Schrader, C. D., Stinner, R. J.: Analysis of the moon surface by neutron-gamma ray inelastic scattering technique. J. Geophys. Res. **66,** 1951 (1961).

— Waggoner, J. A., Berger, J. H., Martin, E. F., Stinner, R. J.: Neutron gamma ray instrumentation for lunar surface composition analysis. ARS J. **32,** 631 (1962).

Sellers, B., Ziegler, C. A.: Radioisotope alpha excited characteristic X-ray sources, Symposium on Low Energy X and Gamma Sources and Applications. Chicago, Illinois, Oct. 1964, Proceedings published by AEC.

Speed, R.: Lunar and planetary X-ray diffraction program. Jet Propulsion Laboratory Memorandum No. **33/218,** 157 (1965).

Spencer, N. W., Reber, C. A.: A mass spectrometer for an aeronomy satellite. W. Priester, ed. Space Research 111: Proc. of thc Third International Space Science Symposium, Interscience Publishers, 1151, 1962.

Trombka, J. I., Metzger, A. E.: Analysis Instrumentation. Plenum Press, New York, 237, 1963.

— Senftle, R. L., Schmadebeck, R. L.: A mobile geochemical laboratory system using remote data analysis and acquisition. Winter Meeting American Nuclear Society, Nov. 30—Dec. 4, San Francisco, Calif., 1969.

— Schmadebeck, R. L.: A numerical least sqare method for resolving complex pulse height spectra. NASA Special Publication 3044, 1968.

Chapter 4: Apollo Surface Missions and the Lunar Receiving Laboratory Program

Among the most intriguing chemical studies by far are those being performed presently on the returned lunar samples. An extraordinary amount of effort and planning has gone into the preparation of a pro-

Table 4.1. *Progression of lunar science*

Ranger-Lunar Orbiter Surveyor	Apollo	Lunar Explorations
Small Scale Features	Initial Manned Landings	Extended Duration Mobile Eploration
Apollo Site Mapping	Field Geology and Photography	Topographic, Geologic, Geophysical Surveys
Science Site Coverage		Detailed Subsurface Investigations
Soft Landing Feasibility	Sample Return for Earth Analysis	— Seismic — Thermal — Electrical
Detailed Surface Texture	First Order — Interior Structure	— Sonic — Radioactive
Soil Strength	— Near Surface Layering — Radiation Environment	Soil Mechanics
Chemical Composition	— Heat Flow	
1964—1968	1969—1970	Sample Collection and In-Situ Analysis Sample Return for Detailed Analysis Remote Sensing of Surface from Orbit Astronomy Observations Erosional Processes Atmospheric Analysis 1971—

gram for dealing with these samples; the return of which constitutes one of the most significant scientific events in the program of space exploration.

The return of lunar samples is an integral part of the second phase of lunar exploration which begins with the Apollo program. The position in time of this program in the evolving scheme is shown in Table 4.1. The principal efforts are being made in three main areas: skilled field investigations by trained astronauts, the return of lunar samples with a carefully recorded geological context, and lastly, the emplacement of geophysical instruments for the long term monitoring of lunar processes.

The analysis of returned lunar samples is the subject of this chapter. The study of these samples is expected to contribute important geochemical, cosmological and biochemical data. Accordingly a very comprehensive series of examinations by some of the world's most competent investigators is taking place.

Landing Sites

The selection of landing sites in the early Apollo missions is highly significant because it is from these areas that samples are returned for analysis and study. The choice of these sites is highly conditioned by topography and the required operational constraints which follow from the performance characteristics of the Apollo vehicle.

The Apollo lunar landing zone consists of a rectangular area extending from plus 45 degrees to minus 45 degrees longitude and from plus 5 degrees to minus 5 degrees latitude (see Figure 4.1). This choice permits earth tracking during the crucial lunar descent and ascent as well as a "free return trajectory".

The five candidate Apollo landing sites are in the form of an ellipse approximately 5 by 7 km. The distribution of these sites is such that at least three sites, about 26 degrees apart in longitude, are available for a landing. One very important consideration in the choice of a site is the nature of the lighting at the time of a landing. This seriously affects the launch window. The objective is to land at a site during a period when the sun angle is somewhere between 7 to 18 degrees.

With regard to the terrain requirements, the early missions require a low regional slope of the order of a few degrees and no large elevation along the approach path.

Keeping the above conditions in mind the five sites selected are distributed as follows; two of the sites are in the sea of Tranquillity, the third site lies in the central bay region and the fourth and fifth sites are in Oceanus Procellarum.

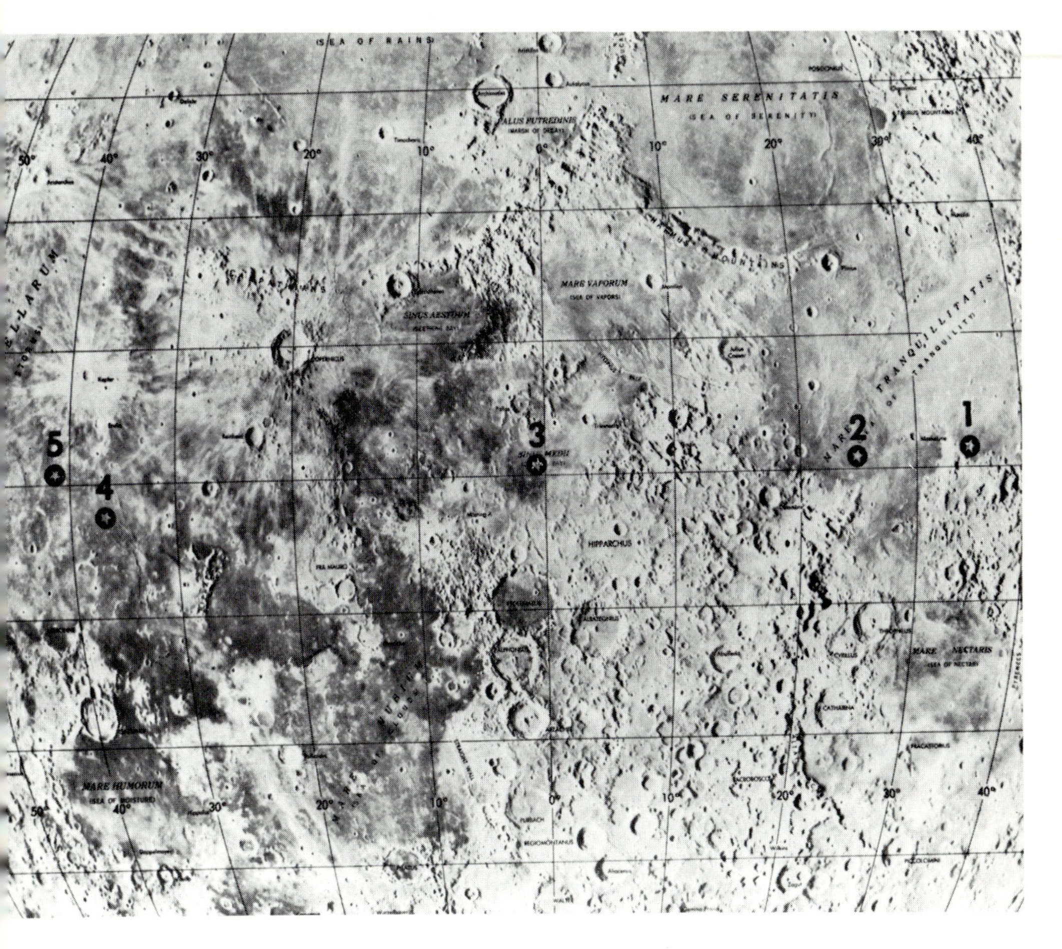

Figure 4.1 Five of the candidate sites selected by the National Aeronautics and Space Administration for the first manned landings. The sites are numbered simply from east to west and lie in a rectangular area between 45 degrees east and west longitude and 5 degrees north and south of the equator

Apollo Mission Profiles

In view of the fact that two Apollo missions have already occurred one can now write the Astronaut's activities on the lunar surface in a historical manner. It is evident that the experience gained will have a large influence on the planning of future missions, and greatly affect the way in which the scientific objectives are implemented.

In the initial stages of planning for the Apollo 11 mission, serious thought was given to two astronaut excursions to the lunar surface involving both a program of sample collection and the emplacement of an extensive series of geophysical experiments, called the Apollo Lunar Surface Experiment Package (ALSEP). In practice it was found necessary to scale down the objectives for the first mission, as dictated by questions of astronaut safety.

The extravehicular activity (EVA) during the Apollo 11 flight was preprogrammed down to the nearest five minutes, involving a large number of functions of a housekeeping nature (equipment checkout) as well as scientific. The astronauts were in constant contact with the earth so that it was possible to record their first hand observations about the lunar terrain.

The astronauts (Armstrong and Aldrin) spent approximately two and one half hours on the lunar surface in the Mare Tranquillitatis, during which time they succeeded in collecting three separate sets of lunar samples totaling approximately 22 kg. These were the contingency, bulk and documented samples. In addition they were successful in deploying a solar wind experiment, and a modified ALSEP consisting of a passive seismometer and a corner cube laser reflector.

The contingency sample was collected almost immediately by the astronaut to guarantee the return of some lunar material in the advent of an aborted mission. This sample of surface material was immediately placed in a teflon bag after being collected about 1.5 m from the Lunar Module. It weighed approximately one kilogram and contained several rock fragments in addition to lunar soil.

The bulk samples and the documented samples were collected by means of a scoop and placed in special containers known as the rock boxes (see Figure 4.3). One container was filled with the bulk and undocumented samples well before the conclusion of the EVA.

The documented samples which were placed in the second rock box consisted of about 20 rocks. These were carefully collected by Armstrong during a 3.5 min period towards the end of the EVA. Two surface cores taken by Aldrin were also included as well as the foil used to collect solar wind particles.

The Apollo 12 mission, the most recent of the lunar landings, was launched in November 1969 to that portion of the Ocean of Storms which served as the landing site of Surveyor 3. This mission successfully met a number of objectives. The astronauts carried out 2 EVA excursions of great scientific significance. They emplaced an ALSEP with a complement of five separate geophysical experiments; a passive seismometer, a solar wind spectrometer, a magnetometer, a suprathermal ion detector, and a cold cathode gauge. At least four of the five instruments are operational and transmitting data. Sample collection was even more successful than during Apollo 11 in that about 80 pounds of sample were collected as compared to the 50 pounds in the first mission. A truly exciting event has been the collection of portions of the Surveyor 3 spacecraft in order to study the effect of extended exposure to the lunar environment.

One significant aspect of these missions is the ability of the suited astronauts to use the specially designed handtools; the hammer, scoop, and coring tool.

Contamination Control

Contamination of the lunar samples during and after collection is a major problem which can make some experiments difficult, if not impossible, to perform. Efforts have been made to minimize contamination, although it is recognized that it is impossible to eliminate it completely. Because of this it was essential to catalog all possible contaminants that might be encountered. To deal with this problem a contamination control program was established with the following objectives:

1. Minimum release of foreign matter on the lunar surface.
2. Control and identification of materials used in the sample tools, the sample containers, the laboratory systems and all components with which the samples are likely to make contact.
3. Precleaning, sterilizing and outgassing of sampling tools, return containers, laboratory systems and similar items.
4. Environmental isolation of samples.
5. Maintenance of an environmental history on each sample in order to identify unavoidable contaminants.

One major source of contamination is the spacecraft engine exhaust during touchdown. Extensive studies have been made of the nature of the contamination as well as it's distribution around the vehicle. A typical distribution map for water is shown in Figure 4.2 as an example (McLane, 1967).

A principal source of biological contamination is the air effluent from the spacecraft and leakage from the rotating joints of the space suit. The cabin effluent is released to the lunar surface through a biological filter. Space suit leakage is a more serious and difficult problem since it represents a moving source of contamination. Tools for sample collection were developed to cope with this problem.

The materials from which the sampling tools and return containers are made have been selected with special care. Since these come into contact with the samples, it has been necessary to minimize or eliminate

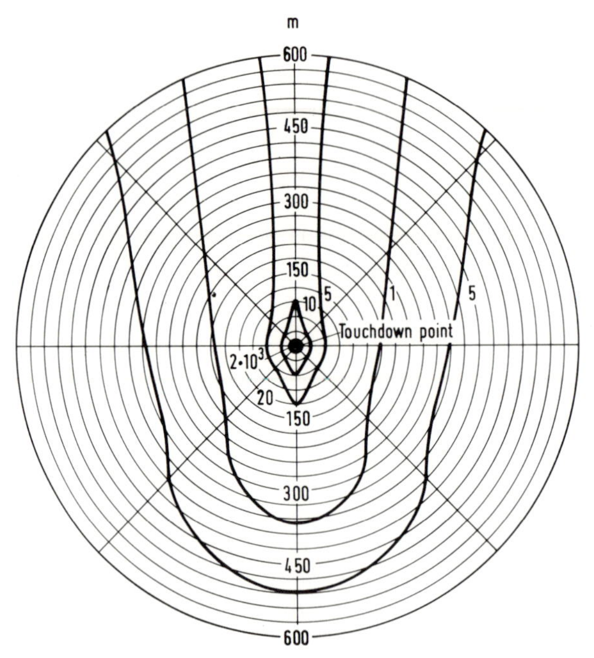

Figure 4.2 Expected lunar surface contamination by water from the LEM (Lunar Excursion Module) descent engine exhaust. The units in the plot are micrograms per square centimeter

those elements of geochemical interest, for example, isotopes used in age dating such as lead, uranium, thorium and potassium. Where elastomers are required only the fluorcarbons such as teflon and Viton A are used because they are distinctly recognizable by mass spectroscopy.

A considerable amount of time and effort has gone into the design of containers for the return of lunar samples. The different types of vacuum containers are shown in Figure 4.3. These consist of a large "rock box" for returning samples at a modest vacuum of about 10^{-6} torr. and a high vacuum container to maintain the ambient pressure at which the sample was collected (approx. 10^{-12} torr.).

Both types of containers are degassed before flight at the Lunar Receiving Laboratory, then sealed before being loaded on the Lunar Module. On the moon the containers are loaded and sealed for the return trip. The large box has a double seal, the outer one an elastomer type and the inner one of crushed indium. If the astronauts have adequate time the collected samples are individually enclosed in teflon bags before being loaded into the large containers. Where time is short or manipulation too difficult, the rock samples are placed directly into the large containers.

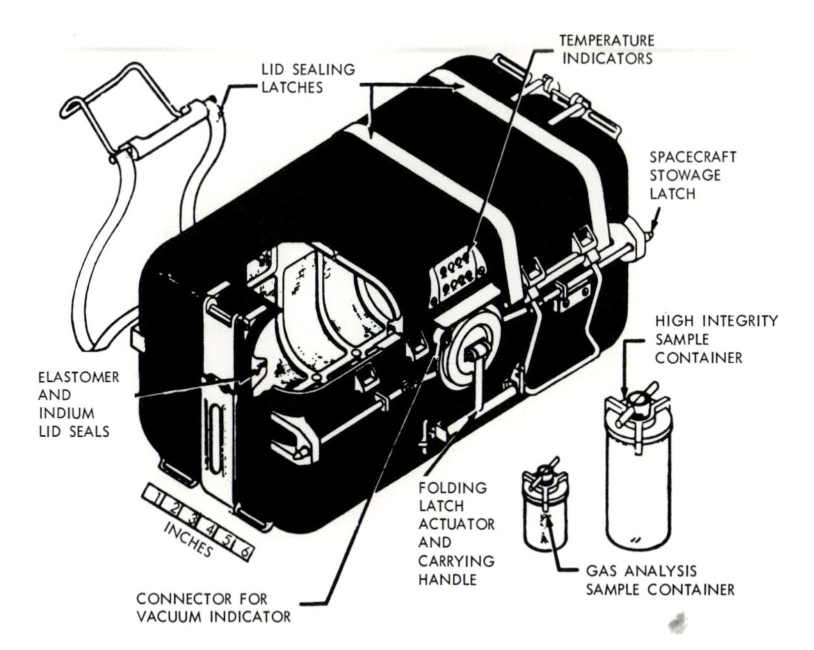

Figure 4.3 Various types of vacuum sample containers for containing the lunar samples after collection. The small containers have been designed to maintain the lunar ambient pressure. After sample storage in the small containers, they are then stored in the larger box for transport

The original plans called for the highly critical biological test samples, as well as those for interstitial and gas analysis, to be loaded into separate, small, high vacuum containers with crushed indium seals which would then be placed inside the larger rock box.

Lunar Receiving Laboratory (LRL)

The Lunar Receiving Laboratory, building 37 at the Manned Spacecraft Center in Houston, Texas, is one of the most unique structures in the world. It's purpose is to provide facilities for performing centralized receiving functions for men and materials returning from extra-terrestrial space missions.

The building is divided into three functional areas:

1. Sample operations Area: This includes the vacuum laboratory, the magnetics laboratory, the gas analysis area, the physical-chemical test area, the biological test laboratories, and the radiation counting laboratory. Except for the radiation counting laboratory, the entire

sample operations area is within the confines of a unit biological barrier system.

2. The Crew Reception Area: This area houses the crew and lies entirely within the biological barrier.
3. The support Laboratory and Administrative Area: This area is not isolated.

The functions of the LRL include the following:

1. Preflight preparations of geologic handtools and sample return containers.
2. Opening and unpacking of the sample return containers.
3. Preliminary physical chemical tests.
4. Sample cataloguing.
5. Biological quarantine clearance tests.
6. Biological isolation of returned men and materials.
7. Time dependent studies.
8. Repackaging and distribution of samples to investigators.
9. Storage of undistributed samples.
10. Sample data storage and retrieval.

Among the various items listed above 5 and 6 are considered as the most vital functions of the LRL. The possibility, however remote, of the existence of viable organisms in any form on the Moon has seriously influenced the design of the LRL as well as the philosophy of its operation. NASA, in cooperation with a number of regulatory government agencies, has undertaken a series of steps to protect the earth's biosphere against the introduction of foreign pathogens. These steps are in accordance with Article IX of the 1967 International Treaty on Principles Governing Activities in the Exploration and Use of Outer Space, Including the moon and Other Celestial Bodies. What is called for is a period of quarantine of both men and materials which have been in contact with the Moon.

The astronauts are generally quarantined for at least three weeks while the lunar samples are kept behind the biological barriers for at least four weeks. During this period, a number of scientists and technicians perform the various tests and functions described below. We shall describe in some detail the various operating components of the LRL. For additional detail the reader is referred to a paper by McLane (1967).

Vacuum Laboratory

The vacuum laboratory functions as a pivotal installation of the LRL in several aspects; the preparation of tools and containers for outbound flight, the reception of the extra-terrestrial samples and finally the distribution of samples to other parts of the LRL and to the scientific community. The vacuum systems of the vacuum laboratory are probably the only ones of their kind in the world, having been developed entirely for the extraordinary function of dealing with lunar samples. The installation occupies two floors; the first floor containing the rough pumps for the vacuum chambers, the electrical distribution center, the instrument power supplies, and the utilities distribution system. The second floor contains the vacuum chambers, the storage carousels, consoles, and conditioning system.

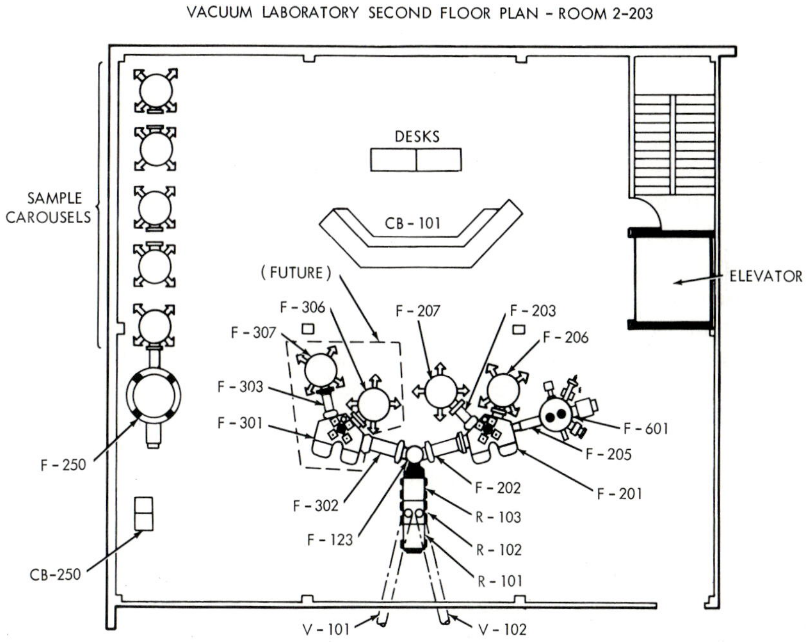

VACUUM LABORATORY SECOND FLOOR PLAN - ROOM 2-203

Figure 4.4 Vacuum Laboratory: V-101,2 are the transfer tubes for physical-chemical test samples, F-201 is the 1×10^{-6} high vacuum glove chamber, F-601 is the 1×10^{-11} ultra high vacuum chamber, F-206 is the tool carousel, F-207 is a sample carousel, F-250 is a conditioning chamber

The arrangement of the vacuum laboratory is shown in Figure 4.4. The chamber, F-250, known as the conditioning chamber is used for preflight preparations such as the outgassing of the Sample Return Containers, as well as high temperature, vacuum sterilization. A similar decontamination function is performed on the various items or tools prior to their storage in the tool or sample carousels. All movement in and out of the conditioning chamber is through vacuum locks. Monitoring and control functions are performed on the control console, CB-250, shown adjacent to the conditioning chamber.

The primary vacuum complex is made up of three divisions:

1. Atmospheric decontamination system.
2. Vacuum transfer locks.
3. Sample processing systems.

The atmospheric decontamination system consists of the cabinets designated as R-101, R-102, R-103, V-101, and V-102. Cabinet R-101, equipped with high intensity ultra violet lamps, serves as an airlock for the vacuum system. R-102, has a peracetic wash system for decontaminating the outer surfaces of the various containers moving through the primary vacuum complex. In addition facilities are provided for a sterile water wash for removing the peracetic acid. In the following cabinet, R-103, a flow of sterile, warm nitrogen is used to dry the containers. V-101 and V-102 are atmospheric transfer locks for use in transferring samples from the vacuum chambers to the bio-prep and physical-chemical cabinets.

The vacuum transfer system consists of three chambers, F-302, F-122, and F-202. These chambers are equipped with an elaborate pumping system consisting of a titanium sublimation pump and a high capacity vac-ion pump backed by a turbo-molecular pump, operating through an absolute biological filter. The details of the pumping system are described in a report prepared by members of the Manned Spacecraft Center at Houston.

The sample processing complex is made up of components labeled F-201, F-205, F-206, and F-207. Figure 4.5 shows the details of chamber F-201 in which the sample containers are opened and processing begun on the lunar samples. The chamber contains a residual gas analyzer, a Cahn 100 gram electro-balance, load cells, six Beattie Coleman automatic cameras; A Leitz binocular microscope, video monitor systems, and assorted instruments to facilitate sampling handling. All operations are to be performed by an operator working through arm gloves into the vacuum chamber. The arm gloves consist of inner and outer teflon gloves with the inner space pumped in order to minimize leakage into the chamber.

Two carousels, F-206 and F-207 (see Figure 4.6) are connected to F-201. One carousel, F-206, is used to store tools and the outer container for the samples moving to the Radiation Counting Laboratory. F-207 is used for storing repackaged lunar samples. Both carousels consist of dual level, lazy susan type storage trays and both are equipped with an extensor monorail system permitting immediate retreival of any item in the carousel. F-203 is a transfer lock for F-207 making it possible to exchange carousels without bringing either the carousel or the lock up to atmosphere.

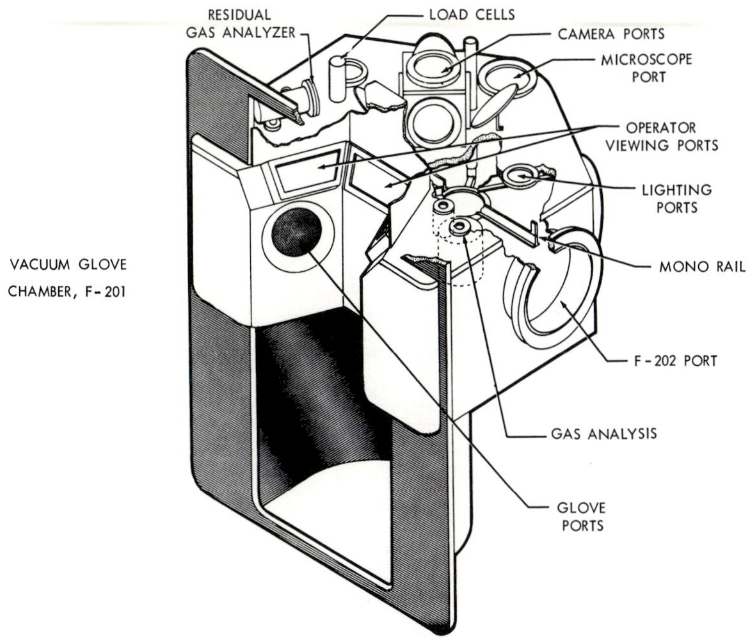

Figure 4.5 High vacuum glove chamber

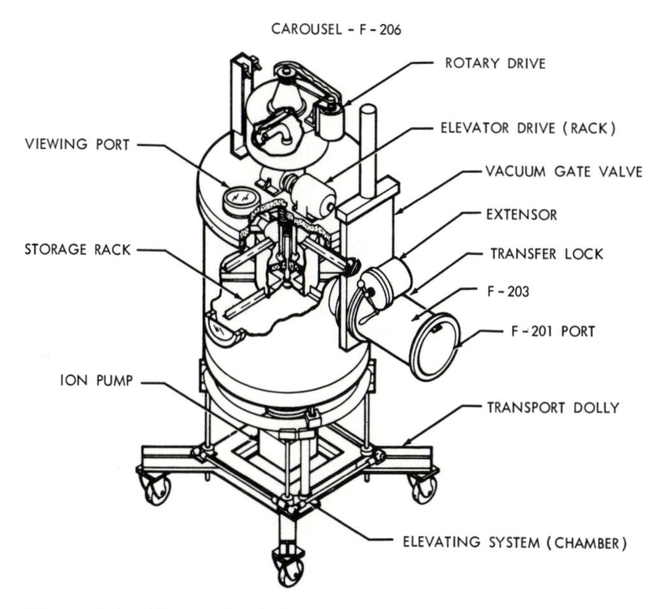

Figure 4.6 Carousel details

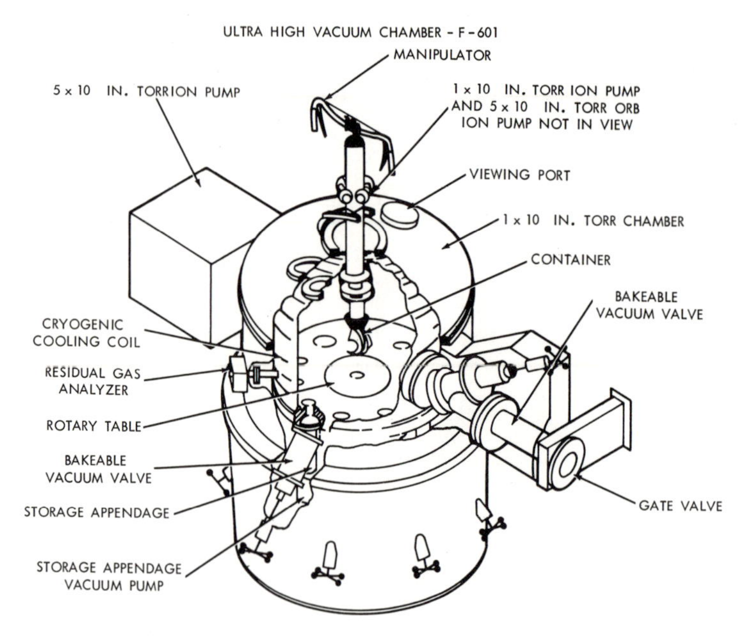

Figure 4.7 Ultra High Vacuum Chamber

Chamber F-601 (Figure 4.7), the ultra high vacuum chamber, is used to subdivide that sample which is to be brought back under lunar ambient pressure. This chamber is a double walled, differentially pumped chamber which operates a pressure of 10^{-11} to 10^{-12} torr. In this chamber a mechanical manipulator is provided in order to minimize leakage. An operator using the manipulator and a rotary table is able to subdivide the samples and store them in individual containers, shown as appendages on the lower left hand side of the main chamber. The arrangement is such that the appendage containers have their own pumping system, thus permitting the shipment of lunar samples outside the laboratory at pressure of about 10^{-11} torr. Once the carousels are filled with lunar samples, they can be disconnected, moved to a storage room and there connected to a pumping system and maintained under high vacuum.

Magnetics Laboratory

The Magnetics laboratory is a specially designed laboratory for performing magnetic measurements on the returned lunar samples. These measurements will be described in more detail below. The laboratory is adjacent to the Vacuum Laboratory, and is carefully shielded by Mu metal in order to maintain a constant and controlled magnetic environment.

Gas Analysis Laboratory

This laboratory is located on a mezzanine above the Vacuum Laboratory on the third floor of the LRL. The layout of the laboratory is shown in Figure 4.8. The facility is equipped with all the necessary instrumentation for effluent gas analysis on the returned lunar samples, both at ambient and elevated temperatures. The ambient temperature measurement is performed on the samples and containers opened in the 10^{-6} torr glove chamber (F-201) previously described. A low resolution, low mass range, and medium sensitivity mass spectrometer, Veeco Model GA-4, is provided for the gas analysis under ambient temperatures. The detector for the apparatus is part of the glove chamber (see Figure 4.5), but the control console is in the gas analysis laboratory.

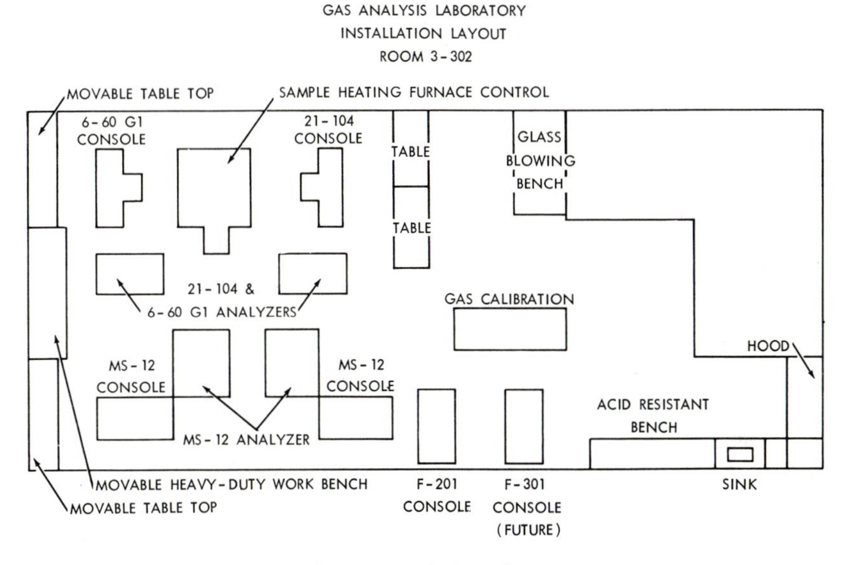

Figure 4.8 Arrangement of the Gas Analysis Laboratory

Two mass spectrometers are provided for inorganic and rare gas analysis. These are a Nuclide Analysis Associates Model 6-60-Gl, which has a medium mass range at high sensitivity, and a Consolidated Electrodynamics Corporation Model 21-104 with similar specifications.

Sample heating is performed with a furnace similar to the Varian Model S-27 which is used with both rare gas and inorganic gas mass spectrometers. The furnaces provide programmable resistance heating from ambient to 600 degrees centigrade and induction heating from 600 degrees to 2000 degrees centigrade. Temperature control is to plus or minus one percent of the indicated values.

Two mass spectrometers are also available for organic analysis. These are high mass range, high sensitivity, medium resolution spectrometers. Readout for real time analysis and display is done with a computer processor.

Physical-Chemical Test Area

This area consists of a chemical laboratory, a spectrographic laboratory, and dark room. Contained in the Physical-Chemical Laboratory is a large number of double sided, gas tight, positively pressured, nitrogen atmosphere biological barrier cabinets with glove ports for two operator positions. Samples move into this cabinet system through a transfer tube from the vacuum laboratory. The chambers in the Physical-Chemical Test Area are highly instrumented with the following items:

1. Cahn electro-balance with a chart recorder.
2. Gas reaction chamber.
3. Gas chromatograph.
4. 2 Leitz binocular microscopes with photographic accessories.
5. Leitz petrographic microscope (Ortho-Lux Pol) with white and monochromatic light sources.
6. Leitz Dia-Lux-Pol microscope with petrographic and Orthomat petrographic accessories.
7. Microchemical test reagents and apparatus.
8. Spectrographic standards and electrodes.
9. Small handtools.

In addition to the items listed above, the laboratory contains casework with chemical and moisture resistant tops and a chemical fume hood. Outside the gas tight cabinetry one finds a petrographic microscope, an Abbe refractometer, a pan balance, refractive index oils, and solid refractive index standards.

The spectrographic laboratory facilities consist of a specially built X-ray fluorescence spectrometer, designed to operate in one of the gas tight cabinets, and a Jarrel Ash Mark IV, 3.4 m, convertible Ebert stigmatic spectrograph. The custom Vari-source Excitation Unit and chamber are also housed in a gas tight cabinet, whereas the spectrograph itself is outside the cabinets. A comparator microphotometer is also part of the spectrographic facility, and is located adjacent to the spectrograph.

Biological Test Laboratory

The biological laboratory arrangement is shown in Figure 4.9. The laboratory is extensively equipped for the performance of various tests

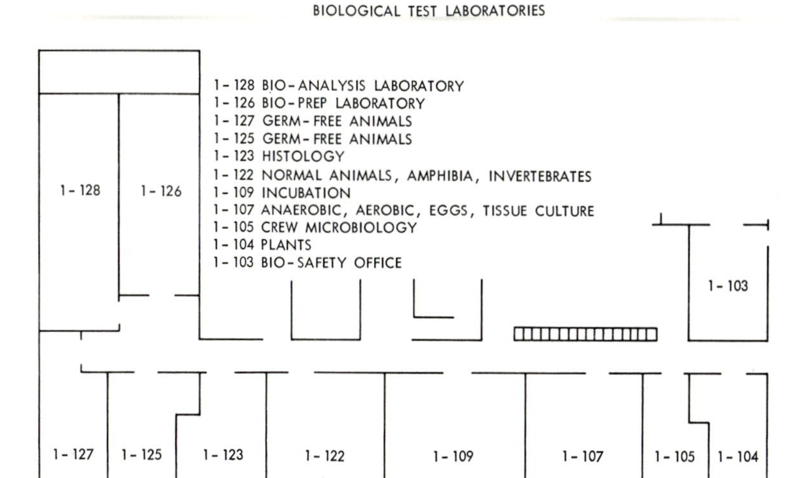

Figure 4.9 The Biological Test Laboratory Complex

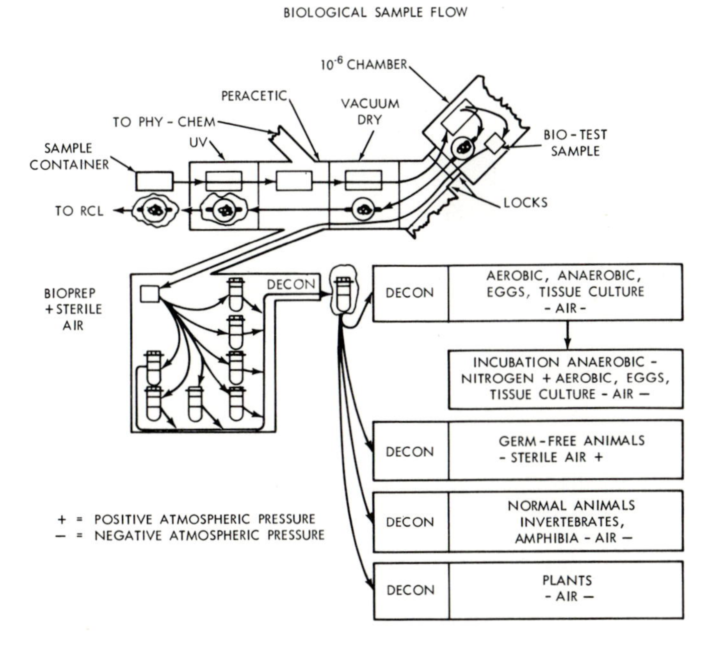

Figure 4.10 The flow of returned lunar samples through the Biological Complex

to determine if the returned samples are in any way a threat to the terrestrial biosphere. All testing is done in biological cabinetry designed to prevent the terrestrial biocontamination of the returned samples, and in turn to prevent contamination of the terrestrial biosphere. The installation consists of a bioprep laboratory, a bioanalysis laboratory, germ free animal rooms, a histology laboratory, a normal animals laboratory, an incubation laboratory, a tissue culture laboratory, a crew microbiology laboratory, and a plant laboratory. The flow of returned samples through the complex is shown in Figure 4.10.

Radiation Counting Laboratory

The counting facility is shown in Figure 4.11. The counting laboratory itself is located underground with offices and support areas above ground. The first floor above ground includes offices for the staff and visiting scientists, and a standards preparation laboratory.

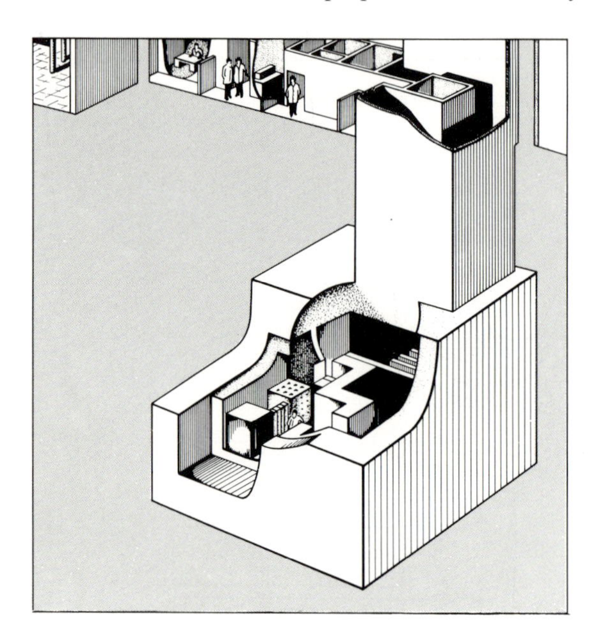

Figure 4.11 The Radiation Counting Laboratory of the LRL is shown here in cutaway. It lies about 40 ft. below the surface

A radon removal system, for removing radon from the gamma ray counting room has been installed on the second floor, above the counting laboratory. The underground portion of the RCL is shown in Figure 4.12. This laboratory is some forty feet below the surface, which effectively eliminates the nucleonic component of cosmic rays and reduces the mu meson flux by about 80% in the counting room. The counting room which is about 13 ft. by 24 ft., is surrounded on all sides by about three feet of dunite (olivine) supported by a steel plate liner. The dunite and all the other materials have been carefully selected because of their low inherent radioactivity.

Air to the counting room is supplied through a radon removal system made up of chilled charcoal beds and absolute filters. The radon removal system is redundant so that continuous operation is possible during regeneration or filter replacement. The steel liner of the counting room is airtight, and entry into the counting laboratory is through an airlock. The room is kept at positive pressure relative to the control room to prevent radon leakage into the counting room.

As shown in Figure 4.12 the counting room contains two internal shields, a main counter shield and an anti-coincidence shield. The main counter shield is composed of lead walls, 8 in. thick, and has an internal volume 7 ft. × 7 ft. × 7 ft. Contained in this volume are two 9 in. × 5 in. sodium iodide crystals. Arrangements are built into the system so that these detectors can be operated at angles other than 180 degrees with respect to each other. The anti-coincidence shield assembly is shown in

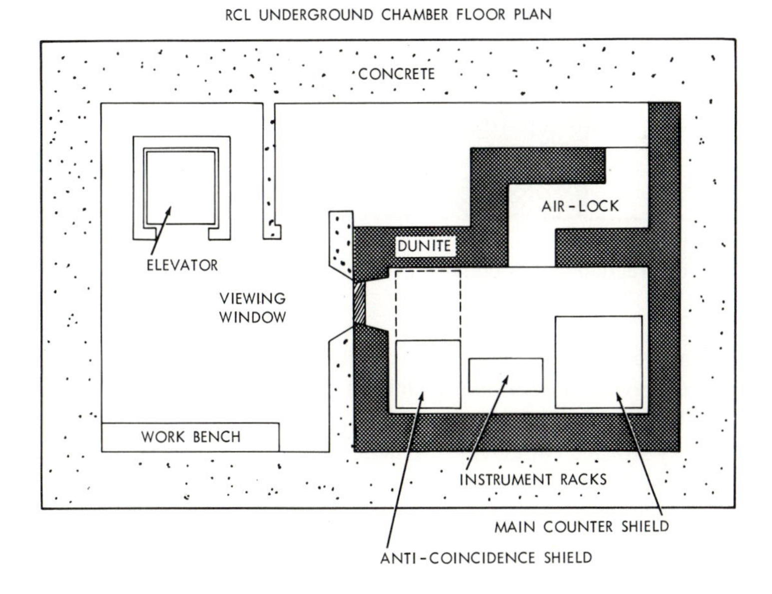

Figure 4.12 The underground chamber floor plan for the Radiation Counting Laboratory

cross section in Figure 4.13. Two 9 in. by 5 in. sodium iodide detectors are oriented at 180 degrees. The detectors are enclosed in a 12 in. thick inner anti-coincidence plastic mantle of polyvinyl toluene. Surrounding the inner mantle is an 8 in. thick lithiated lead shield and an outer coincidence mantle. The inner mantle improves the peak to total ratio by rejecting events in the principle detectors where only a partial conversion of the gamma ray energy has occurred. The outer mantle is sensitive only to meson interactions. The passive lead shield contains a small concentration of lithium in order to shorten the lifetime of meson produced neutrons. The effect is to reduce the required length of the anti-coincidence inhibit pulse.

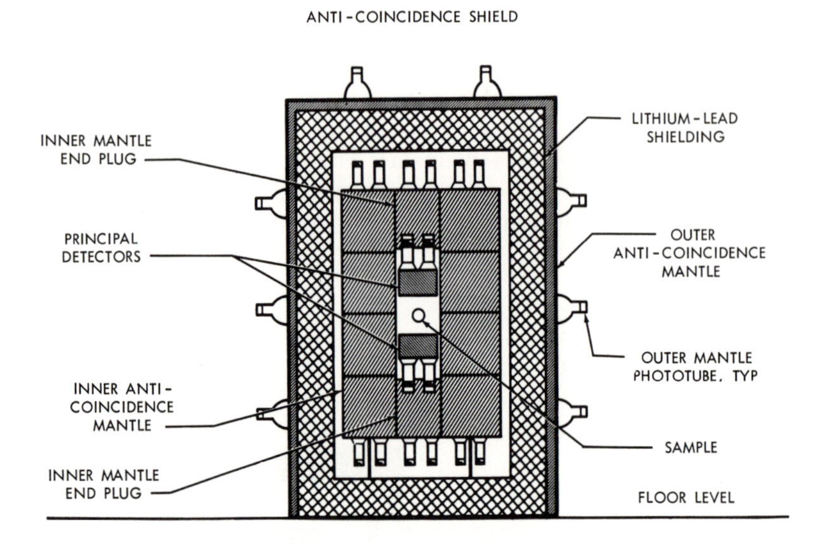

Figure 4.13 Construction details on the anti-coincidence shield

Data from the detectors is acquired with a multi-parameter pulse height analyzer having the following characteristics:

1. 4096 ADC (analog to digital converter) with 12 bit resolution.
2. Fast memory storage capacity equivalent to 128 by 128 channels.
3. Buffer storage for writing unsorted addresses appropriate to each single or coincidence event onto magnetic tape at an ADC resolution of 7—12 bits.
4. A small general purpose computer to permit data from two independent, two parameter analyzers to be multiplexed into the system with each event tagged as to origin.

Biological Support Laboratory

These laboratories provide for the various needs of the biological areas within the sample operations area. The facilities include a pouring room for the vitro culture media, a media preparations area, a room for the cleaning and sterilizing of glass ware, a safety office for monitoring and controlling the biological containment facilities, a histological laboratory, a tissue culture laboratory, an animal suite, etc. Included among the biological support laboratories is an electron microscope suite with a high resolution electron microscope for examining lunar samples as well as material isolated from the samples. Sample preparation is done in one of the cabinets of the Sample Operations Area, and the sample grids for the electron microscope are then transported to the instrument.

Sample Flow Through the Lunar Receiving Laboratory (LRL)

As indicated above the function of the LRL begins prior to the actual Apollo flights. The Apollo Lunar Sample Return Container (ALSRC) is prepared for the outbound flight at the LRL. This preparation involves extensive cleaning, degassing, sterilization, sealing and leak checking. At the completion of these operations the containers are delivered in protective covering to the launch site.

On return from the moon the containers are packaged aboard the recovery vessel in biological isolation containers. On delivery to the LRL the sample containers are immediately taken to the vacuum laboratory through an airlock. Once in the vacuum laboratory processing begins.

A number of investigators work in the LRL during the quarantine period. These are some of the Principal Investigators that ultimately perform more detailed studies. These scientists have worked at the LRL for a period of time prior to the receipt of samples for training and coordination with the LRL activities and for calibration and testing of the LRL equipment. These LRL scientists supply data to the investigators not working at the LRL to help in providing background information and better interpretation when the detailed studies are performed.

Once the sample containers are in the LRL they are brought into the primary vacuum complex for opening and preliminary examination of the lunar samples they contain.

Detailed sampling handling procedures for the samples are under constant study and revision. Figure 4.14 shows the nature of the sample flow through the LRL. The program has been developed with very great emphasis on detail and represents the joint efforts of the LRL personnel under the direction of P. R. Bell as well as a committee consisting of scientists that have been selected as principal investigators. The details of these procedures are not within the scope of this book but they can be found in numerous documents published by the Manned Spacecraft Center of NASA. We shall however describe below the preliminary examination and the physical measurements performed.

Radiation Counting

In step 3 (Figure 4.14) we observe that samples have been removed for measurement. These consist of a general sample and three special samples. Since the general sample is destined for immediate radiation counting to yield information on short lived radionuclides, cosmic ray produced, it must be processed with great speed. The plans are to accomplish the necessary preliminary examination as quickly as possible.

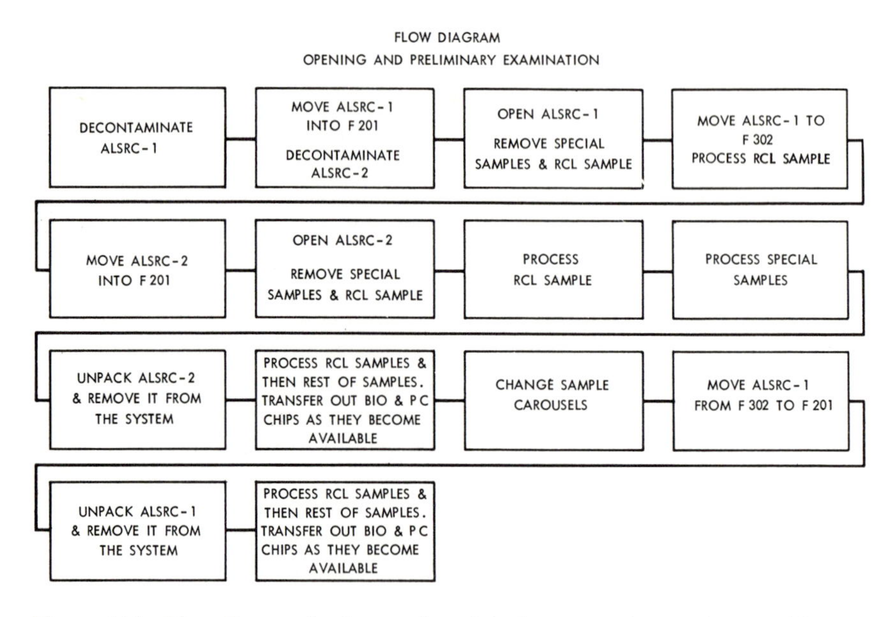

Figure 4.14 Flow diagram for the opening of the lunar sample containers and for the preliminary examination

This examination involves photographing the sample from six different views, removing biological and physical-chemical chips, taking a scan of the sample and finally once more taking photographs from six views. The sample is weighed again, examined microscopically and finally hermetically sealed in a teflon blag and stored in the sample carousel until an appropriate container is fabricated as described below:

The strategy involved in the radiation counting consists of sample preparation, transfer to the counting laboratory, counting, storage, recounting, return to the vacuum laboratory, preparation of standards, counting of standards and data reduction.

Gamma ray analysis of irregularly shaped samples requires standards of the same size and shape, of comparable density and radioactivity as the sample. It is also essential to position and orient the samples and standards in an identical manner. This requirement is met by fabricating an inner counting container and support ring to hold the sample and then the standard during the respective counting periods. In the particular case of lunar sample analysis, because of biological and vacuum constraints, the sample, inner counting container and support ring are to be sealed inside a spheroidal outer counting container. The procedure for fabricating the inner container is shown in the following figure (Figure 4.15).

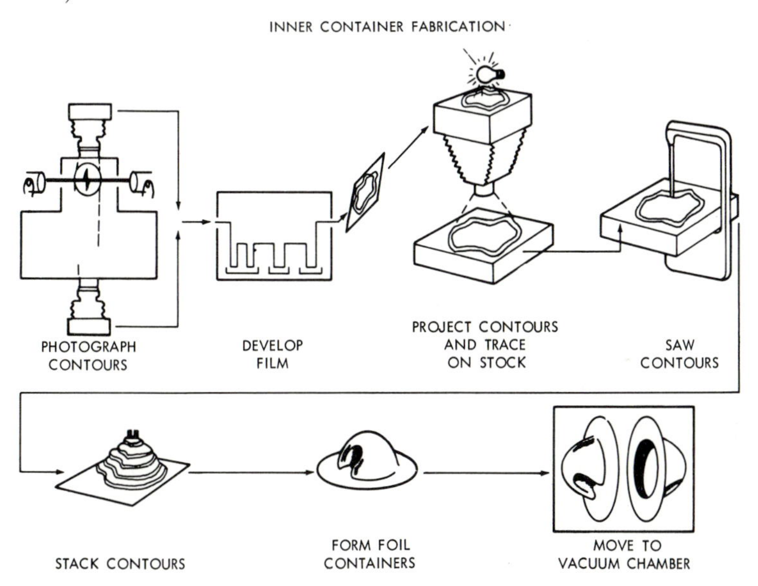

Figure 4.15 Procedure for fabricating the inner container for the radiation counting procedure

The present procedure calls for the first sample to be ready for counting in about 7 h after the arrival of the sample container at the LRL. Subsequent samples will require about 2 to 5 h.

Radiation standards are prepared by mixing rock and metal powders with radioisotopes, simulating the density and radioactivity of each lunar sample.

Finally the data, recorded on magnetic tape is reduced by computer techniques using programs and library spectra specifically developed for the RCL equipment.

Two types of gas analysis are performed: 1. The analysis of gas evolved at ambient temperatures from the lunar samples and their containers as they are opened in the vacuum glove box and 2. the analysis of occluded and interstitial gases which are generated at elevated temperatures. Figure 4.16 is a schematic representation of the glove chamber gas handling and collection system for the ambient temperature gas. As already described above this system is used to analyze the gases in the Apollo Lunar Sample Receiving Outer Containers and the individually sealed field sample bags containing the general purpose lunar samples. The various operations involved are performed manually at the glove chamber while in voice contact with the mass spectrometer console operator stationed in the Gas Analysis Laboratory. A typical procedure to be performed is shown in Figure 4.17. It can be seen that the sample path depends on the nominal pressure observed in the bag. The cold traps shown as T_2 through T_9 in the previous figure are liquid nitrogen adsorbate types. These traps are attached to a removable manifold so that if necessary a new manifold may be attached for addi-

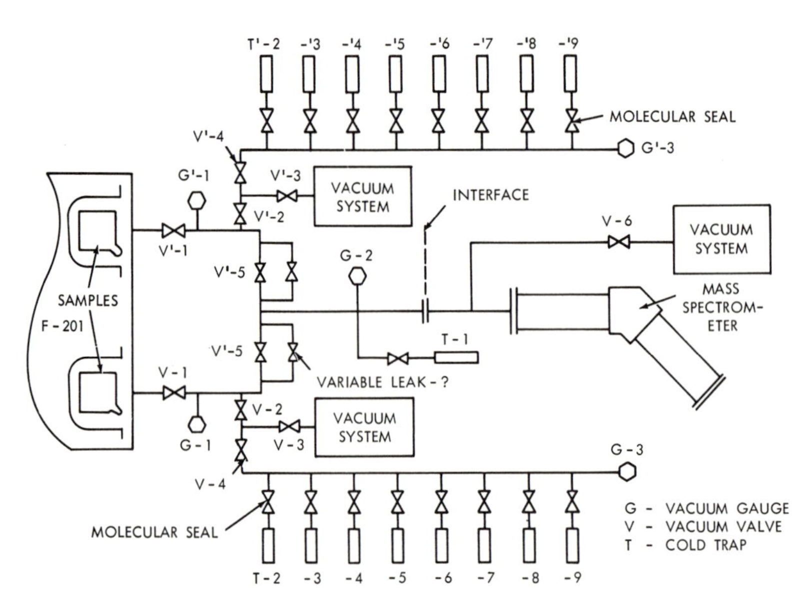

Figure 4.16 Schematic diagram of the glove chamber gas handling and collection system

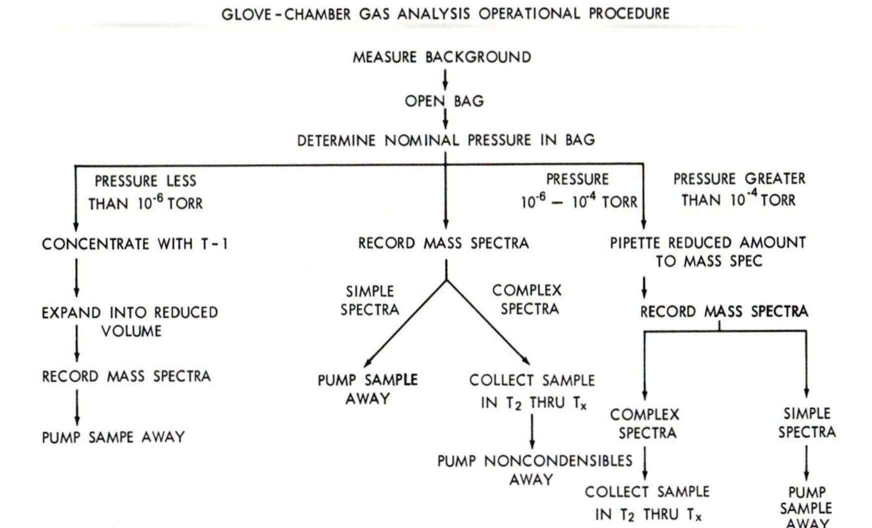

Figure 4.17 Glove Chamber gas analysis procedure

tional sample collection. The samples of gas trapped in this arrangement are transported to the gas analysis laboratory for detailed analysis on the large mass spectrometers.

The Gas Analysis Sample Container provides special lunar samples for the analysis of occluded and interstitial gases. These are evolved by heating of the sample. The sample is carried in a transfer container to the Gas Analysis Laboratory and attached to an opening device on the sample heating furnace. Analysis are conducted for inorganic gases, rare gases, organic volatiles and organic pyrolysis products as a function of temperature. The procedures followed, while still tentative can be described as follows: The effluent gases from the furnace are to be run into an extraction system in order to separate the inorganic gases, rare gases and organic gases. The inorganic and rare gases are passed into the high sensitivity, low resolution mass spectrometer for qualitative and quantitative determinations.

The organic gas components are analyzed by two different mass spectrometric techniques. In one the sample effluent is passed directly into the inlet system of a medium resolution mass spectrometer with a mass resolution between 3000 and 4000, by passing a sample resevoir. The connecting lines are kept at temperatures of 300 degrees centigrade or higher. In order to minimize surface effects the connecting lines are coated internally with ceramic or other inert material.

A complete mass scan can be performed in a few seconds with immediate repeat runs. The plans call for variations in ionizing voltage in order to allow for group type and molecular weight determinations. Each run will be recorded and referenced with respect to time and temperature.

The second method will combine gas chromatography with mass spectrometry, including a molecular separator to increase the sample concentration in the elluent stream of the chromatograph. The mass spectra of each of the components coming from the gas chromatographic column will be observed. The function of the gas chromatograph is for preselection which should result in simplified mass spectra and more reliable qualitative and quantitative data.

Preliminary Physical-Chemical Tests

The small chips or splits of each lunar sample taken initially are used for preliminary mineralogical, petrological and geochemical examination. These samples as we have already indicated come from the Vacuum Laboratory where they are weighed and sealed in vacuum containers at 10^{-6} torr. or better prior to their delivery to the Physical-Chemical Laboratory. The containers are opened in a dry nitrogen atmosphere while the sample weight and the gas near the sample is monitored. The sample chips are continuously observed through a binocular microscope

Requirements

- The protocol is expected to cover the following major items as required for the evaluation of the possible effects of the lunar samples upon the terrestrial biosphere:
  - Biological analysis
  - Life detection
  - Elemental anylysis
  - Containement
- Insure rational program
- Insure compatibility with apollo program
- Satisfy requirements of interagency committee
- Provide firm equipment and staffing requirements

Constraints

- Thirty day maximum sample containment
- Containment and protocol may be extended if definite signs of pathogenicity
- Sample can only be released on contingency basis
- Crew quarntine to be minimized—21–30 days unless definite signs of pathogenicity
- Must be carried within present lunar laboratory design

Figure 4.18 Protocol for sample bio-analysis

to observe if any significant changes occur. Similar reaction studies are done under oxygen and terrestrial atmospheres at various controlled humidities. Should no significant reaction occur, the sample is then moved to an adjacent cabinet for optical mineralogy and petrology studies, observation being performed using binocular and petrographic microscopes. Selected portions of the chips undergo microchemical qualitative and quantitative analysis. A portion of the chips are also passed into the spectrographic preparation cabinet where they are prepared, loaded into electrodes and analyzed by emission spectroscopy for major and minor elements.

Magnetic Monopole Experiment

One particularly unique study performed at the LRL is the search for magnetic monopoles in the lunar materials. The procedure followed involves spinning the samples in vacuum tight containers in cryogenically cooled coils and searching for small induced currents.

Biological Quarantine Testing

The primary objective of the biological testing is to determine if the returned lunar samples in anyway represent a threat to terrestrial life. In order to perform this function a large number of studies are performed involving such disciplines as microbiology, virology, animal and plant pathology/histology and bacteriology. The protocol for sample bio-analysis is shown in Figure 4.18. The results of these studies determine when and if the samples are released to the scientific community for more detailed studies.

Sample Distribution

Following the preliminary examinations and the time dependent studies at the LRL and after the quarantine period, the lunar samples are distributed to previously approved investigators. The amounts and sample state are wherever possible, those specified by the investigators. Some samples are in the form of chips or powder. In other instances the specimens distributed include petrographic thin sections and polished thin sections, as well as samples maintained under high vacuum or inert atmospheres.

Detailed Studies of Lunar Samples

The previous sections have described the functions of the LRL and the testing to be performed during the quarantine period. We have seen that such functions include biological quarantine tests, preliminary visual examinations, repackaging for storage and for shipment, gas analysis and time dependent studies such as radiation measurements and some sample preparation such as the making of thin sections for study at the LRL and for distribution. After the quarantine period and if the samples are demonstrated to be free of hazards to terrestrial life, they are then distributed to a number of preselected scientists for further and more detailed studies.

At this writing over 180 principal investigators have been carefully selected by the NASA office of Space Science and Applications to perform a great variety of experiments. The selection has been made on the basis of a thorough evaluation by a number of sub-committees appointed by NASA, the subcommittees themselves consisting of a number of prominent scientists. Approval has been contingent on a number of factors such as the scientific merit of the submitted proposal, the competence and reputation of the scientist and the nature of the sample requirements. Included among these investigators are American scientists as well as a substantial number of foreign scientists representing England, Germany, Canada, Japan, Finland, Switzerland, and Australia. Table 4.2 lists the participating universities and institutions. The bulk of these are the universities although there are a substantial number of research institutes and a small number of companies.

A partial list of the proposed investigations shows the large scope of studies to be done. Unquestionably these will be the most thoroughly analyzed samples known to man. The list follows below:

Table 4.2. *Partial List of Participating Organizations*

American Universities

California, Berkely
California, Los Angeles
California, San Diego
California Institute of Technology
Chicago
Cornell
Harvard
Houston
Minnesota
Kentucky
Massachusetts Institute
of Technology
Miami
Missouri
Oregon
Oregon State
Pittsburgh
Princeton
Rice
State University of New York
at Stony Brook
Washington
Wisconsin
Yale

American Research Institutes

Ames Research Center, NASA
Argonne National Laboratories
Batelle Memorial Institute
Brookhaven National Laboratories
Carnegie Institute of Technology
Lamont Geological Observatory
Lawrence Radiation Laboratories
Manned Spacecraft Center, NASA
Oak Ridge National Laboratories
Smithsonian Institution,

Graduate Research Center
of the Southwest
Goddard Space Flight Center, NASA
Illinois Institute of Technology
Jet Propulsion Laboratory
Astrophysical Observatory
Smithsonian Institution,
National Museum
US Geological Survey

American Companies
Douglas Aircraft Company
General Electric Company
Mobil Research and
Development Corp.

Foreign Universities
Australian National University
University of Berne
University of Bristol
University of Cambridge
University of Cologne
University of Durham
University of Edinburgh
University Libre De Bruxelles
University of Manchester
Newcastle Upon Tyne
Queen Mary College
Royal Holloway College
University of Sheffield
University of Tokyo
University of Tübingen

Foreign Research Institutions
Atomic Energy Research
Establishment, Harwell England
Geological Survey of Canada
Geological Survey of Finland
Max Planck, Heidelberg
Max Planck Institut für Chemie,
Mainz

Mineralogy and Petrology

1. Determination of microstructure characteristics and composition.
2. Determinative mineralogy for opaque minerals, distribution of radioactive materials by autoradiography, analysis for lead, uranium, and thorium isotopes by mass spectrometry.
3. Standard petrology.
4. Determination of structure, compositon, and texture of opaque minerals.
5. Determination of pyroxene content by X-ray and optical methods.
6. High pressure/temperature phase studies and determination of the temperature of crystallization of the minerals.
7. X-ray and microprobe analysis.
8. Examination for condensed sublimates.
9. Microprobe analysis and study of diamonds if present.
10. Determination of temperature of rock formation by study of plagioclase properties.

Crystallography

1. Crystal structure of sulfides.
2. Measurement of structural defects through study of optical, electrical, and mechanical properties.

3. Mossbauer and NMR techniques to measure the oxidation state of iron, radiation damage and the energy states of aluminum, sodium and iron in the crystals.
4. Crystal structure and stabilities of the feldspars.
5. Determination of the valence states and symmetry of the crystalline material using electron spin and nuclear magnetic resonance techniques and spin relaxation studies.

Electronmicroprobe studies

1. Elemental analysis.
2. Mineral phase studies.

Radiation effects

1. Effect of cosmic radiation-study of fossil tracks produced by charged particles.

Shock effects

1. Petrographic, microscopic and X-ray diffraction studies of shock on minerals and rocks.
2. Replication and thin section electron microscopy.

Alpha Particle-Autoradiography

1. Identification of alpha emitting nuclides by autoradiography and alpha particle spectroscopy.

Chemical and Isotopic Analysis

1. Wet chemical analysis for major elements.
2. Trace elements by emission spectroscopy.
3. X-ray fluorescence spectroscopy elemental analysis.

Neutron Activation

1. Elemental abundances.
2. Analysis for major rock forming elements.
3. Determination of rare earth content.
4. Determination of the low abundance isotopes of potassium, calcium vanadium and chromium.
4. Trace element determination.
5. Neutron activation analysis for uranium, thorium, bismuth, lead, thallium and mercury.
6. Determinations for gallium, germanium and boron.

Mass Spectroscopy

1. Isotopic abundances of heavy elements.
2. Determination of the concentrations of the alkali, alkaline earths, and lathanide elements.
3. Analysis for rubidium, strontium and their isotopes.
4. Determination of the rare gases, potassium and uranium and identification of cosmic ray produced nuclides.
5. Determination of lead isotopes, concentrations of uranium, thorium, lead and their occurrences in minerals in the lunar samples.
6. Determination of argon, rubidium, strontium and rare gases such as helium, neon, krypton and xenon.

Cosmic Ray Induced and Natural Activity

1. Measurement of cosmic ray produced ^{26}Al.
2. Determination of cosmic ray and solar particle activation effects.
3. Non destructive gamma ray spectrometry for cosmic ray and natural radionuclides.
4. Determination of ^{22}Na, and ^{54}Mn by gamma ray spectroscopy.
5. Measurement of cosmic ray induced radioactive nuclides ^{14}C and ^{36}Cl.

Light Stable Isotopes

1. Determination of the stable isotopes of oxygen, carbon, hydrogen, silicon, deuterium, and nitrogen.

Physical Properties

1. Search for magnetic monopoles.
2. Measurement of magnetic properties.
3. Remanent magnetism studies.
4. Identification of magnetic minerals.
5. Thermo-magnetic and magnetic susceptibility studies.
6. Measurements of sonic velocity, thermal expansivity, specific heat, dielectric constant and index of refraction.
7. Measurements of electrical properties and thermal conductivity.
8. Particle size analysis.
9. Photometric studies of radiation effects from several types of rays.
10. Calorimetry (thermal properties).
11. Measurement of elastic-mechanical properties.

Bioscience Program

1. Mass spectrometric analysis for organic matter in the lunar crust.
2. Analysis of organic lunar samples for alpha amino acids and polymers thereof.

3. Identification of organic compounds in the lunar samples by gas chromatography-mass spectroscopy, nuclear magnetic resonance, high speed liquid chromatography and variants thereof.
4. A hunt for the presence or absence of lipids, amino acids and polymer type organic matter.
5. Analysis of lunar material for $^{13}C/^{12}C$ and deuterium/hydrogen ratios in organic matter.
6. Electron microscope studies.
7. Detection of viable organisms by growth.

Results of the Preliminary Examination of the Apollo 11 Lunar Samples

The above sections have described in detail the various activities leading to the acquisition and study of lunar samples. Now we shall summarize the actual results obtained on the Apollo 11 lunar samples. The examination was performed by the Lunar Sample Preliminary Examination Team (LSPET) at the Lunar Receiving Laboratory in Houston. The team was supported by other scientists and contract personnel normally attached to the LRL. The number of individuals involved was substantial and the combined results were issued as a joint report in Science (1969)*.

---

* The people who contributed directly to obtaining the data and/or to the preparation of the Science report are: D.H. Anderson, Manned Spacecraft Center (MSC); E.E. Anderson, Brown and Root-Northrop (BRN); K. Beeman (MIT); P.R. Bell (MSC), D.D. Bogard (MSC); R. Brett (MSC); A.L. Burlingame, U. California, Berkeley; W.D. Carrier (MSC); E.C.T. Chao, US Geological Survey (USGS); N.C. Costes, Marshall Space Flight Center; D.H. Dahlem (USGS); G.B. Dalrymple (USGS); R. Doell (USGS); J.S. Eldridge, Oak Ridge National Laboratory (ORNL); M.S. Favaro, US Dept. of Agriculture (USDA); D.A. Flory (MSC); C. Frondel, Harvard Univ.; R. Fryxell, Washington State Univ.; J. Funkhouser, State University of New York, Stoneybrook; P.W. Gast, Columbia University; W.R. Greenwood (MSC); M. Grolier (USGS); C.S. Gromme (USGS); G.H. Heiken (MSC); W.N. Hess (MSC); P.H. Johnson (BRN); R. Johnson, Ames Research Center; E.A. King (MSC); N. Mancuso (MIT); J.D. Menzies (USDA); J.K. Mitchell, Univ. of Calif., Berkeley; D.A. Morrison (MSC); R. Murphy (MIT); G.D. O'Kelley (ORNL); G.G. Schaber (USGS); A. Schaeffer, Univ. of N.Y. Stoneybrook; D. Schleicher (USGS); H.H. Schmitt (MSC); E. Schonfeld (MSC) J.W. Schopf, Univ. of Calif. at L.A.; R.F. Scott, Calif. Institute of Technology (CIT); E.M. Shoemaker (CIT); B.R. Simoneit, Univ. of Calif., Berkeley; D.H. Smith, Univ. of Calif., Berkeley; R.L. Smith (USGS); R.L. Sutton (USGS); S.R. Taylor, Australian National University; F.C. Walls, Univ. of Calif., Berkeley; J. Warner (MSC); R.E. Wilcox (USGS); V.R. Wilmarth (MSC); and J. Zähringer, Max-Planck-Institut, Heidelberg, Germany.

Geologic Description

A brief description of the geologic setting of the landing site as detailed in the report is as follows: The Apollo 11 spacecraft landed in the southwestern part of Mare Tranquillitatis at 0.67° N and 23.49° E, about 10 km southwest of the crater Sabine D. The landing site is crossed by faint but distinct north-northwest trending rays associated with the crater Theophilus, 320 km to the southeast. Another more prominent north-northeast trending ray lies 15 km west of the landing area. Because the landing site lies between major rays, the possibility exists that it contains some fragments coming from Theophilus, Alfraganus, Tycho and other distant craters.

A sharp rimmed crater was observed about 400 m east of the landing point. This crater which is approximately 180 m in diameter and 30 m deep, has been named West crater. It is surrounded by a blocky ejecta apron, extending outward for about 250 m from the rim crest. Large blocks some as large as 5 m across are found on the rim and the crater interior. In addition rays of blocky ejecta (0.5 to 2 m across) extend beyond the ejecta apron west of the landing point.

The lunar surface at the landing point consists of unsorted fragmental debris ranging from particles too fine to be resolved by the naked eye to blocks 0.8 m across.

The mare surface near the Lunar Module contains numerous small craters ranging in diameter from a few centimeters to tens of meters. All of the craters near the LM have rims, walls and floors of relatively fine grained material with scattered coarser fragments occurring with the same abundance as in the inter-crater areas. The coarse fragments scattered in the vicinity of the LM are presumed to come from the nearby blocky rim crater.

Description of the rocks

The rock fragments at the Apollo 11 site were found to have a variety of shapes. Most of the rocks were embedded in the fine material of the regolith to varying degrees. The rocks were either rounded or partially rounded on their upper surfaces although angular, irregular fragments were also observed. A few of the rocks were rectangular slabs with a faint platy structure. Of great interest was the fact that many of the rocks, rounded on their exposed surfaces, were flat or of irregular shape on the bottom. It has been proposed that the shape of these rocks suggests that erosional processes are occurring on the lunar surface which produce a gradual rounding of the exposed rock surfaces. Some rounded rock surfaces actually show individual clasts and grains as well as glassy lined pits on their surfaces in raised relief as a result of the wearing away or ablation of the surface. The rocks exhibiting this differential erosion most clearly are the microbreccia. The ablation is thought to be due to small particle bombardment of the rock surfaces. An additional mechanism that has been proposed to account for the observed rounding is the peeling away of closely spaced exfoliation shells. This may explain the rounding of some crystalline rocks of medium grain size.

The rounded upper surfaces of most of the rocks show minute, deep pits from a fraction of a millimeter to 2 millimeters in diameter. Many of these pits are lined with glass. The pits differ in appearance from impact craters produced in the laboratory and their origin is as yet unknown.

One very interesting observation made by astronaut Armstrong was the presence of blebs of material at the bottom of 6 or 8 of the raised rim craters. These blebs were 2 to 10 cm wide, with specular surfaces. They appeared to be glassy, resembling drops of solder and suggestive of splashing molten material impacting at low velocities. It is anticipated that samples returned in the Apollo 12 sample collection will shed some light on these materials.

Mineralogy and Petrology

The returned samples have been divided into four groups:
1. Type A, fine grained vesicular crystalline igneous rock.
2. Type B, medium grained vuggy crystalline igneous rock.
3. Type C, breccia.
4. Type D, fines.

The Science report uses the term “rocks” for fragments greater than 1 centimeter in diameter. Fragments less than 1 centimeter are labeled “fines”. The rocks as well as the smaller fragments are described as showing “unearthly surface feature” (see Figure 4.19). Clearly shown are the unusual glassy surface pits in the breccia.

The crystalline rocks are best described as volcanic in appearance but the Preliminary Examination Team has pointed out that this implies surface lava or near surface igneous rock. No conclusion has been drawn on whether this has resulted from impact produced volcanic action as opposed to terrestrial type volcanism.

The crystalline rocks contain pyrogenic assemblages and gas cavities and appear to have crystallized from melts. A number of the phases which have been identified are those usually associated with rock forming minerals, with some exceptions. Some of the minerals which have

Figure 4.19 Breccia showing clearly defined glassy pits

already been identified are clinopyroxene, plagioclase, ilmenite, troilite, cristobalite, olivine and metallic iron with and without appreciable nickel. The crystalline rocks have an unusually high concentration of opaque minerals (ilmenite) compared to terrestrial olivine bearing basalts. Noteworthy is the complete absence of hydrous mineral phases.

The breccias are mixtures of fragments of different rock types and are gray to dark gray in color with specks of white, light gray and brownish gray fragments. Most of the breccias are fine grained containing fragments smaller than one centimeter in diameter and mostly under 0.5 cm. The fragments making up the breccia consist of the types of rocks or minerals described above. In addition there are angular fragments and glassy spherules showing a wide range of color and refractive indices. Glassy spherules are also found in the fines.

Two types of unique surface features have been observed on all the rocks. These are small pits lined with glass and glass spatters. The pit diameters are generally less than one millimeter. Besides the glassy pits there are also thin glass crusts, perhaps due to spattering. Figure 4.20 is a striking example of the types of glassy spherules found in the lunar fine material.

Two core samples were also returned. These were each 2 cm in diameter, one 10 cm long and the other 13.5 cm long. These cores contained particles with diameters ranging from 1 mm to 30 microns and consisted

Figure 4.20 Glassy spherules found in the lunar material

of angular rock fragments, crystal fragments, glass spherules, aggregates of glass and lithic fragments. On opening the coring tools, the material was found to maintain a cylindrical shape. Analysis showed the fines to be mainly a variety of glasses and the minerals clinopyroxene, ilmenite and olivine. In addition rare spherules and rounded fragments of Ni—Fe were observed.

Chemical Analysis

The chemical analysis performed at the LRL was of a preliminary nature. The principal technique employed was optical spectrography with a quoted precision of $\pm 10\%$. The accuracy of the results was controlled by using a number of standard samples such as a set of international rock standards supplied by the US Geological Survey (these cover a fairly complete spectrum of rock types from ultra-basic to granitic), analyzed terrestrial basalts from Hawaii and Galapagos, chondritic meteorites (Forrest City and Leedey) and achondrites (Sioux City and Johnstown). Three samples were sterilized and brought out of the biological barrier for atomic absorption analysis. Table 4.3 shows the abundances of some of the major and minor elements expressed as oxides for 12 of the lunar samples, including rocks, breccia and the fines. Table 4.4 lists the abundances of the trace elements in the same samples.

It is worth noting that the samples were found to be free of inorganic contamination from either the rock box or the rocket exhausts from the lunar module (the skirts of the rocket engine would have been expected to contribute a substantial amount of niobium as a contaminant).

The preliminary chemical analysis has contributed the following interesting facts:

1. There is a marked similarity in composition of the various samples. The more significant variations are shown by some of the trace and minor elements such as Ni, Zr, Rb, and K. The major elemental constituents are Si, Al, Ti, Fe, Ca, and Mg. The minor elements are Na, Cr, Mn, K, and Zr.
2. The lunar rocks show an unusually high concentration of refractory elements such as Ti, Zr, and Y. A comparison with chondritic meteorites shows the lunar material to have higher concentrations of Ca and Al but lower concentrations of Mg and Fe.
3. Zr, Sr, Ba, Y, and Yb are enriched by about 2 orders of magnitude in the lunar samples as compared to chondritic meteorites.
4. K and Rb are present in lunar rocks in amounts similar to those found in chondrites whereas Ni and Co are depleted. Ni was not detected in some of the rocks (Less than 1 ppm.) although the Fe remained high.
5. The concentration of Zr is unusually high in the lunar rocks.
6. The ratios of Rb to Sr are low, similar to those found in terrestrial oceanic basalts. Ba, Cr, and Sc are relatively abundant.
7. The volatile elements such as Pb, Bi, and Tl, if present, are below the limit of detection of the spectrographic technique. This is true as well for the elements of the Pt group and Au and Ag.

Summarizing the above it can be stated that the lunar rocks, on detailed analysis, do show differences from terrestrial or meteoritic samples.

Table 4.3. *Elements detected. Abundances expressed as weight in percent oxides*

	Type A rocks (vesicular)				Type B rocks (crystalline)				Type C rocks (breccias)		Type D fine material	Bulk Bio-Pool sample
Oxides	22*	72	57	20	17	58	45	50	21	61	37	54
SiO_2	43	45	36	38	40	43	42	38	43	40	43	42
Al_2O_3	7.7	9	11	11	10	13	13	11	11	12	13	13
TiO_2	11	10	12.5	12	11	9	8	9	8.6	10	7	7.0
FeO	21	17	20	18	19	17	18	20	19	16	16	15.6
MgO	6.5	8	9.5	8	8.5	6.5	7	10	7.4	9	8	7.6
CaO	9.0	9.5	10	10	10	10.5	10	10	11	11	12	11.6
Na_2O	0.40	0.60	0.54	0.59	0.65	0.56	0.51	0.51	0.20	0.48	0.54	0.50
K_2O	0.21	0.20	0.18	0.064	0.22	0.11	0.10	0.064	0.15	0.17	0.12	0.14
MnO	0.26	0.36	0.49	0.32	0.35	0.55	0.27	0.50	0.22	0.41	0.23	0.34
Cr_2O_3	0.41	0.69	0.95	0.31	0.67	0.54	0.51	0.70	0.37	0.69	0.37	0.41
ZrO_2	0.14	0.11	>0.27	0.13	0.19	0.03	0.095	0.095	0.20	0.04	0.05	0.07
NiO	0.04	**	**	**	**	**	**	0.007	0.03	0.04	0.03	0.015
Total	99.0	100.5	101.4	97.8	100.5	100.8	99.5	99.9	99.8	99.8	100.3	98.2

* Laboratory number. ** No data.

Table 4.4. *Elements detected. Elemental abundances*

	Type A rocks (vesicular)				Type B rocks (crystalline)				Type C rocks (breccias)		Type D fine material	Bulk Bio-Pool sample
Element	22*	72	57	20	17	58	45	50	21	61	37	54
Rb (ppm)	**	6.5	6.0	1.5	6.0	1.6	1.9	0.8	**	3.1	2.2	2
Ba (ppm)	100	130	180	50	120	85	115	60	105	90	68	65
K (percent)	0.17	0.17	0.15	0.053	0.18	0.09	0.084	0.053	0.12	0.15	0.10	0.11
Sr (ppm)	110	55	230	85	55	190	60	140	150	60	90	140
Ca (percent)	6.4	6.8	7.1	7.1	7.1	7.5	7.1	7.1	7.9	7.9	8.6	8.3
Na (percent)	0.30	0.44	0.40	0.44	0.48	0.41	0.38	0.38	0.15	0.37	0.40	0.38
Yb (ppm)	7	2	6	2.5	**	5	1.3	2.7	4.5	1.8	2.5	2.5
Y (ppm)	230	210	310	185	310	230	100	130	300	115	130	200
Zr (ppm)	1000	850	>2000	980	1250	250	700	700	1500	400	400	500
Cr (ppm)	2800	4700	6500	2100	4600	3700	3500	4800	2500	3000	2500	2800
V (ppm)	36	30	40	20	30	32	40	80	22	32	42	30
Sc (ppm)	110	45	110	110	55	130	90	170	68	55	55	60
Ti (percent)	6.6	6.0	7.5	7.2	6.6	5.4	4.8	5.4	5.2	5.4	4.2	4.2
Ni (ppm)	320	***	25	***	***	***	***	55	215	235	250	120
Co (ppm)	15	12	22	3	10	7	7	10	13	12	18	11
Cu (ppm)	**	5	**	4.5	3	**	6	10	**	8	**	**
Fe (percent)	16	13	15.5	14	14.7	13	14	15.5	14.8	12.4	12.4	12.1
Mn (ppm)	2000	2800	3800	2460	2700	4300	2100	3900	1700	2400	1750	2600

*Laboratory number. ** No data. *** Not detected.

Rare Gas Analysis

The rare gas analysis was performed at the LRL in a facility specifically constructed for this purpose. Samples were prepared under nitrogen and under air. Rock chips were weighed, wrapped in aluminum foil and sterilized at 125—150 °C for periods from 5 to 24 h. Observations showed that these procedures released less than 1 % of the gas present.

In the analysis the samples were melted by radiofrequency induction heating in a molybdenum crucible. The released gases were purified with a hot titanium getter. The heavier noble gases were condensed and all the gases measured by mass spectrometry. The measurements were made on three fractions: He and Ne, Ar and Kr, and Xe.

The rare gases in the lunar samples show three patterns which relate to the three rock types; the breccia, igneous rocks and fines. The breccia and fines contain extremely high concentrations of rare gases, particularly in the surface and subsurface material. The isotope ratios and amounts of rare gas lead to the conclusion that the solar wind is the primary source of the gases. The igneous rocks show substantially smaller quantities of the rare gases, due either to losses during formation or that the igneous rocks represent subsurface material. Measurements on temperature release from the fines and breccia show the noble gases to be tightly bound rather than surface adsorbed.

The results of the rare gas analysis are summarized in Table 4.5. Based on these measurements the Preliminary Examination Team has listed the following observations:

1. The evidence shows the accumulation of large amounts of rare gases of solar composition.
2. There is an enrichment of neon-20 relative to neon-22, predictable on the basis of the nuclear processes occurring in the sun.
3. The ratio of helium-4 to helium-3 in the soil and breccia is approximately 2600 in accordance with theoretical estimates.

Table 4.6 shows the isotopic ratios for Xe in the fines and breccia. The isotopic pattern is similar to that of trapped Xe in carbonaceous chondrites except for a small amount of the lighter Xe isotopes, (due to spallation reactions) and a defficiency of ^{134}Xe and ^{136}Xe. The ratio of $^{129}Xe/^{132}Xe$ is similar to that found in carbonaceous chondrites.

Preliminary Age Determinations

Because several of the crystalline rocks contain radiogenetic ^{40}Ar spallation produced noble gases, and potassium, it has been possible to do a K—Ar dating measurement and a determination of cosmic ray

Table 4.5 *Rare gas analysis*

Sample type	Helium		Neon			Argon			Krypton	Xenon
	10^{-8} cc/g He^4	4/3	10^{-8} cc/g Ne^{20}	20/22	22/21	10^{-8} cc/g A^{40}	40/36	36/38	10^{-8} cc/g	10^{-8} cc/g
Lunar soil										
Sample A	10,900,000	2500	200,000	13	31	39,000	1.1	5.3	38	38
Sample B	19,000,000	2500	310,000	13	30	42,000	1.2	5.4	36	16
Typical agglomerate rock										
Sample A	15,100,000	2900	320,000	13	29	150,000	2.3	5.3	73	46
Sample B	16,000,000	2700	230,000	13	29	110,000	2.2	5.2	49	42
Typical crystalline rock										
Sample A	63,000	180	210	3.1	1.3	5700	96	1.2	0.34	0.65
Sample B	28,000	270	200	7.1	2.3	1600	42	2.4	0.19	0.16

Table 4.6 $^{129}Xe/^{132}Xe$ *ratios*

Sample	124	126	128	129	130	131	134	136
Fines	0.0062	0.0071	0.086	1.07	0.165	0.829	0.373	0.306
Breccia	0.0052	0.0057	0.084	1.07	0.164	0.820	0.370	0.304

exposure. Seven rocks examined yielded consistent ages of $3.0 \pm 0.7 \times 10^9$ years. Radiation exposure ages vary from 10×10^6 years to approximately 160×10^6 years.

Gamma-Ray Measurements

Gamma ray measurements were performed on eight lunar samples in the Radiation Counting Laboratory. Among these are a sample of fines, a rock from the bulk sample box, one rock from the contingency sample and five rocks from the documented sample box.

Because of the newness of the operation and the complex operation, counting was not begun in the RCL for about four days after the samples reached the LRL. Thus radioactive species with short half lives (less than four days) became undetectable. Additionally the high activity from the Th and U and daughter products caused considerable interference in the determination of the weak gamma ray components. A compilation of the results obtained are tabulated in Table 4.7. These are summarized below:

1. Twelve radioactive species were identified, some tentatively. The shortest half lived species observed were ^{52}Mn (5.7 days) and ^{48}V (16.1 days).
2. The concentrations listed in Table 4.6 represent whole sample averages.
3. The K concentration is variable and near to that for chondrites (0.085 weight percent). The U and Th are near the values for terrestrial basalts and the ratio of U to Th is about 4.1. One remarkable difference is that the ratio of K to U is unusually low, much lower than similar ratios for terrestrial rocks and meteorites.
4. Cosmogenic ^{26}Al is generally high. The ^{26}Al measurements in accord with the rare gas analysis indicate a cosmic ray exposure of several million years.

Organic Analysis

A survey of organic constituents by a pyrolysis-flame ionization detector method and by means of a very sensitive mass spectrometer, provided an estimate of the indigenous organic content of the lunar samples. The values published give the organic content as under 10 parts per million.

Conclusions

The major findings, based on the preliminary examination are listed below. A more complete picture of the lunar samples must await the more detailed studies which are presently under way.

Table 4.7. *Summary of gamma-ray analyses of lunar samples*

Characteristic	Sample number							
	571	721	030	170	181	191	211	026
Weight (g)	897	399	213**	971	213**	245**	216**	302
Geologic type	A	A	B	B	C	C	C	D
K (wt., %***)	0.242±0.036	0.232±0.035	0.050±0.008	0.227±0.034	0.144±0.022	0.12±0.02	0.120±0.018	0.11±0.02
Th (ppm)	3.4±0.7	2.9±0.4	0.95±0.14	2.9±0.4	2.3±0.3	1.9±0.3	1.8±0.3	1.6±0.3
U (ppm)	0.78±0.16	0.75±0.11	0.20±0.03	0.70±0.10	0.60±0.09	0.43±0.06	0.39±0.06	0.46+0.10
^{26}Al (dpm/kg)	77±16	70±15	69±14	66±13	100±20	98±20	81±16	97±19
^{22}Na (dpm/kg)	44±9	42±9	41±8	34±7	55±11	47±10	41±8	44±9
^{44}Ti (dpm/kg)	*TI*	*TI*	–	–	–	–	–	*TI*
^{46}Sc (dpm/kg)	10±5	13±4	13±3	11±3	13±4	10±4	10±4	9±3
^{48}V (dpm/kg)	–	–	*TI*	*TI*	–	–	–	–
^{52}Mn (dpm/kg)	–	–	39±18	–	–	–	36±20	–
^{54}Mn (dpm/kg)	40±13	20±8	26±5	38±13	28±14	27±14	15±7	28±9
^{56}Co (dpm/kg)	30±12	30±10	38±6	18±6	33±11	35±11	38±13	27±9
^{7}Be (dpm/kg)	*TI*	–	*TI*	*TI*	–	–	–	*TI*

* Values for short-lived nuclides have been corrected for decay to 1200 h, c.d.t., July 21, 1969.

** Weight certain; see text.

*** K determined by arraying ^{40}K and assuming terrestrial isotopic ratios for potassium.

1. The mineralogy and texture of the rocks divides them into two genetic groups; fine and medium grained crystalline rocks of igneous origin and breccias.
2. The crystalline rocks differ from terrestrial rocks and meteorites.
3. The appearance of the rocks suggest a strong erosional process different from terrestrial processes.
4. Chemistry suggests that the crystalline rocks were formed under highly reducing conditions (low partial pressures of O_2, H_2O and S).
5. The absence of secondary hydrated minerals suggests the absence of surface water at the landing site during any part of the rocks exposure.
6. There is evidence of shock or impact.
7. The rocks show glass lined surface pits, due perhaps to impact by small particles.
8. The fines and the breccia contain large amounts of noble gases. The elemental and isotopic data indicate the solar wind as the source.
9. K—Ar dates indicate crystallization ages of about 3×10^9 to 4×10^9 years.
10. Indigenous organic matter is very low (less than 1 ppm).
11. The chemical composition of rocks and fines is similar.
12. There is an enrichment of some refractory elements such as Ti and Zr and a depletion of the alkali and some volatile elements.
13. Elements normally enriched in iron meteorites (Ni, Co, and Pt group) were either absent or in very low abundance.
14. The ratio of K to U is unusually low as compared to terrestrial rocks.
15. High ^{26}Al concentration indicates long cosmic ray exposures.
16. There is no evidence of biological matter.

References

McLane, J. C. jr.: Collecting and Processing Samples of the Moon, Astronautics and Aeronautics, 34—47, May, 1967.

Lunar Receiving Laboratory, Building 37, Apollo Missions, Preliminary Report, Manned Spacecraft Center, Houston, Texas, 1966.

Lunar Receiving Laboratory, MSC Building 37, Facility Description, Preliminary Report, Manned Spacecraft Center, Houston, Texas, 1966.

Preliminary Examination of Lunar Samples from Apollo 11, Science, vol. 165, No. 3899, Sept. 19, 1969.

Chapter 5: Data Processing and Analysis

Introduction

A key to proper performance of the various experimental techniques for remote analysis is the proper accumulation, transmission, and analysis of spectral data. Of necessity, measurements are made in a field environment where it is very difficult to exert as much control over the experimental procedures as is usual in a laboratory environment. Thus one must develop data acquisition, transmission, and analysis methods in which the maximum amount of information can be obtained.

As space programs for lunar surface and planetary exploration move from the early exploration phase to more detailed investigative phases, more sophisticated instrumental systems (some of which have been discussed in earlier chapters) are being developed to perform such investigations as surface elemental analysis. Instrumental systems are designed, for example, to aid the astronaut in sample selection, and to be used in obtaining geochemical maps of planetary surfaces.

In order to accomplish the above goals, the authors have evolved a functioning systems approach to remote elemental analysis. The system, as developed, can be divided into five categories:

1. Excitation sources and detectors.
2. Data preprocessor systems.
3. Data transmission systems.
4. On line data analysis systems.
5. Return and display of analyzed data.

By dividing the system in this way, portable, low power, and rugged instrumentation can be taken into the field. Reasonably detailed analyses can be obtained because of direct access through communication links to large on-line computers.

The excitation sources and detectors have already been discussed. Data transmission systems depend on the particular vehicle and mission, and a general discussion of this aspect is beyond the scope of this study. The emphasis in this chapter will be on general analytic methods and how they can be applied to on-line data analysis.

Pulse Height and Digital Spectra

In many of the measurements we have considered, naturally occurring or induced particulate or photon radiation has been detected. It is from the spectral nature and intensity of the measured flux that compositional information is inferred. The measured spectra, however, are not necessarily identical with the true energy spectra required to perform an analysis. The response of a detector to a delta function energy spectrum is not a delta function spectrum, but is rather a characteristic response function which follows from the energy imparted to the detector via the various interaction mechanisms characteristic of the given detector. In the discussion that follows we will refer to the measured spectrum as the digital or pulse height spectrum. The detector output is usually an analog voltage or current signal which is normally converted to digital information by means of an analog-to-digital converter (as, for example, in a pulse height analyzer).

We will attempt to develop a general analytic method for dealing with the digital spectrum. The method to be described will use X-ray and gamma ray spectra for illustration, although the procedure is completely general and has been applied to problems in alpha back-scatter, gas chromatography, and mass spectroscopy.

Before proceeding with detailed discussion of the analytic technique, we shall restate the problem for emphasis. Briefly, the question is whether one can infer the nature of an environment from a measured spectrum resulting from the interaction of that environment with a detector (if all one knows is the characteristic response of the detector). It is assumed, of course, that these response functions of the detector have been determined as functions of the various parameters to be encountered in the course of a measurement. The objective, then, is to obtain results independent of the detector systems (i.e., remove distortions introduced by the detector system). There are different methods for determining the response functions; they can be derived by analytical or empirical methods, and described as a numerical array.

To begin, we can write an expression which relates the digital (pulse height spectrum) and the "true" or energy spectrum:

$$Y(V) = \int_0^{E_{max}} T(E) S(E, V) dE, \tag{1}$$

where $T(E)$ is the "true" or differential energy spectrum as a function of energy, E, $Y(V)$ is the pulse height spectrum as a function of pulse height, V, and $S(E, V)$ is the detector interaction function which reflects the mechanism by which the energy, E, is converted into a pulse height spectrum, V. The desired information is the differential energy spectrum, $T(E)$.

Measurement of Pulse Height Spectra

In order to obtain a solution for $T(E)$ in equation (1) the measurement of $Y(V)$ must be performed correctly. Generally the spectrum of $Y(V)$ is not measured as a continuous function, but as discrete samples because of the finite increment capabilities of a real analyzer (in practice, there are only a finite number of energy channels or storage units). For example, a multi-channel analyzer measures all pulse heights in ΔV about V. One must select a ΔV consistent with the finite resolution of the detector system so that the analyzer itself produces no significant information loss in $Y(V)$.

The following method derived from information theory techniques can be used to determine the ΔV (Trombka, 1962). We assume that the energy resolution of the detector can be described by a Gaussian distribution:

(2) $$Y(V) = e^{-x^2/2\sigma^2}$$

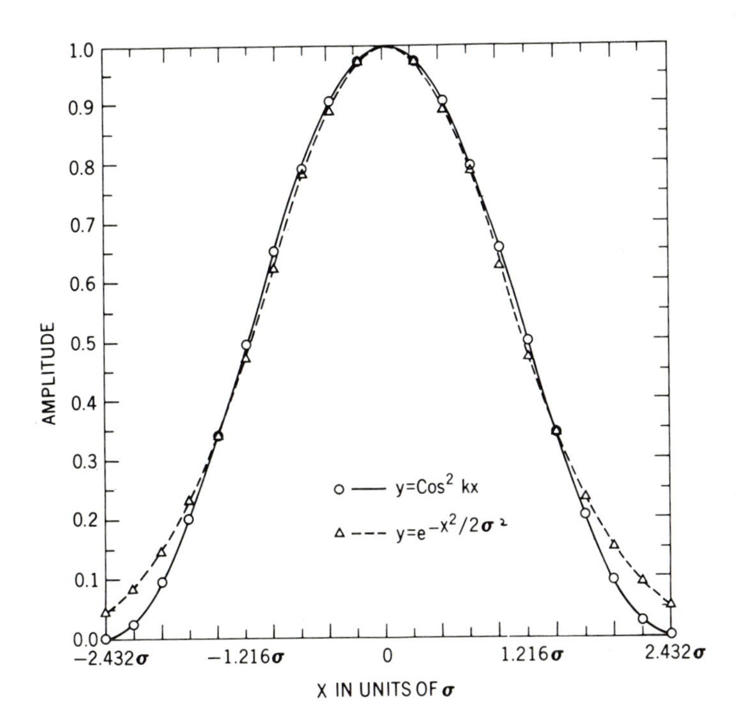

Figure 5.1 Comparison of Gaussian distribution with a $\cos^2 kx$ distribution. Fitted at $x=0$ and $x=\sqrt{2}\sigma$

where $x=(V-V_0)$, and V_0 is the pulse height corresponding to the true incident energy, E_0, and σ^2 is the variance characteristic of the system resolution.

A discrete line, $T(E)=E_0$, will be measured as a pulse height spectrum, $Y(V)$, in the way described in equation (2), assuming that the spectrum given by $T(E)$ has been normalized to correspond to the above equation. Now if a distribution has no oscillatory component with a frequency greater than some f_{max}, where f_{max} corresponds to the frequency of oscillation of the most rapidly changing portion of the measured distribution, then the Shannon Sampling Theorem asserts that samples at discrete points not further than $\frac{1}{2} f_{max}$ describe the original function exactly. In fact, the original function can be reconstructed from the samples (Linden, 1959). We now make a further assumption that $Y(V)$ in equation (2) is negligible and/or equal to zero for those values of pulse height where $Y(V)=10^{-3}$ or less (this assumption does not affect the analysis, because such values of $Y(V)$ are lost in the noise or background for most detectors).

A final assumption is that the Gaussian equation (2) can be approximated closely by a $\cos^2 kx$ distribution; i.e., the amplitudes of the higher frequencies are negligible in terms of the analysis (see Figure 5.1).

The frequency of oscillation is determined in the following manner: The value of k is chosen so that for $x=0$ and for $x=\sqrt{2}\sigma_m$ (i.e. the e^{-1} point on the Gaussian), the amplitude of the Gaussian and $\cos^2 kx$ function are equal; k is found to be equal to $\pi/4.85\sigma_m$. The frequency, f_m, can be determined by the following argument:

$$\cos^2 kx = \tfrac{1}{2}(1+\cos 2kx), \tag{3}$$

therefore,

$$f_m = \frac{2k}{\pi} = \frac{1}{2.43\sigma_m}. \tag{4}$$

It is now possible to relate this frequency to the total width at half maximum ($W_{\frac{1}{2}}$) of the Gaussian distribution. That is, at $Y=\frac{1}{2}$, $X=W_{\frac{1}{2}}$, and $\sigma_m = W_{\frac{1}{2}}/2.35$. Substituting this value of σ_m into equation (4), the maximum frequency, f_m, is found in terms of $W_{\frac{1}{2}}$ to be

$$f_m = \frac{1}{1.03\,W_{\frac{1}{2}}}. \tag{5}$$

For all practical purposes, therefore, the maximum frequency is inversely proportional to the total width at half maximum. We can now consider how this information is applied, taking as an example gamma ray spectroscopy.

The energy resolution, R, of a NaI (Tl) detector is usually quoted for the 0.661 MeV line of a ^{137}Cs. This resolution is defined as:

$$R = \frac{W_{\frac{1}{2}}}{E}, \tag{6}$$

where $W_{\frac{1}{2}}$ is the total width at half maximum, and E is the energy of the given monoenergetic line. This measurement is independent of the use of either pulse height or energy. R is unitless as long as $W_{\frac{1}{2}}$ and E have consistent units. As an example, the resolution for a NaI (Tl) detector can be written as a function of energy as follows:

$$R = \alpha E^{-n}, \tag{7}$$

where α is a constant of proportionality, and n is a constant characteristic of the given detector.

Let us assume that $n=\frac{1}{3}$ (this is a reasonable assumption for a $3'' \times 3''$ NaI (Tl) crystal over the energy range of 0.100 MeV to 3 Mev based on our experimental work). We further assume that our detector has an energy resolution of 8% for the 0.661 MeV line of ^{137}Cs. We now ask, how wide a channel increment must be used in order to measure the pulse-height spectrum for the energy region extending from 0.100 MeV to 1 MeV? The objective, as we have stated, is not to degrade the information beyond that caused by the finite energy resolution of the detection system. The maximum frequency criteria equation (5) will be combined with the minimum value of $W_{\frac{1}{2}}$ in the distribution to be measured. In this instance it will be associated with the minimum energy of interest, i.e. 0.100 MeV. The resolution at 0.100 MeV ($R_{0.100}$) can be calculated from the resolution at 0.661 MeV ($R_{0.661}$) using equation (7). That is:

$$R_{0.100} = R_{0.661} \left(\tfrac{0.661}{0.100}\right)^{\frac{1}{3}}, \tag{8}$$

or

$$R_{0.100} = 15\% \quad \text{for} \quad R_{0.661} = 8\%.$$

From equation (6),

$$W_{\frac{1}{2}} = 0.015 \text{ MeV}. \tag{9}$$

Shannon's Sampling Theorem states that the measurements, (ΔI), should be no farther apart than:

$$\Delta I = \frac{1}{2} f_m = \frac{W_{\frac{1}{2}}}{2}. \tag{10}$$

In our particular problem, $I = 0.0075$ MeV. Because we are interested in measuring the spectrum up to 1 MeV, the number of channels, N_c, of equal width required to perform the measurement would be:

$$N_c = \frac{1 \text{ MeV}}{0.0075} \simeq 133, \tag{11}$$

to the nearest total channel required for an analyzer having equal channel widths. One can decrease the number of channels required by increasing the channel width corresponding to the increase in $W_{\frac{1}{2}}$ as a function of energy. This technique is rather difficult to use, however; therefore equal channel widths are used.

We have thus considered how to determine $Y(V)$ without degrading the information through the use of a pulse height analyzer. We now return to the basic problem of determining $T(E)$ in equation (1).

Formulation of the Least-Square Analysis Method

In most of the cases considered here, analytic solutions for $T(E)$ in equation (1) cannot be obtained; it becomes necessary to develop numerical methods to obtain solutions. The particular method of solution is strongly dependent on the function, $S(E, V)$. A generalized discussion of the functions to be solved is greatly dependent on the detectors used. Thus, two detectors will be considered in order to understand the nature of the function.

We shall first consider detection of monoenergetic X-rays with a proportional counter.

Take the case of an ideal spectrometer looking at an FeK spectrum containing a number of monochromatic energies. We would observe a spectral distribution consisting of a number of discrete lines as shown in Figure 5.2. If, in practice, the X-ray energies are examined by means of a proportional counter detector and a pulse height analyzer, the spectrum becomes a continuous distribution as shown in Figure 5.2. Here we see the effect of instrumental smearing caused by such factors as resolution, statistical fluctuation in the detection process, detector geometry, electronic noise, etc. In many cases the spectrum is further complicated by an escape peak phenomenon resulting from partial absorption of the incident energies.

In the X-ray case the distribution shown does not represent the detector response to a monoenergetic emitter, but represents a response to radiation from a single element source which may in fact be made up of a number of discrete energies. The analytical method to be discussed can deal with either case. In every instance the distribution as shown will be known as the standard, or library, function.

A much more complex function is that characteristic of the response of a NaI (Tl) detector to gamma rays.

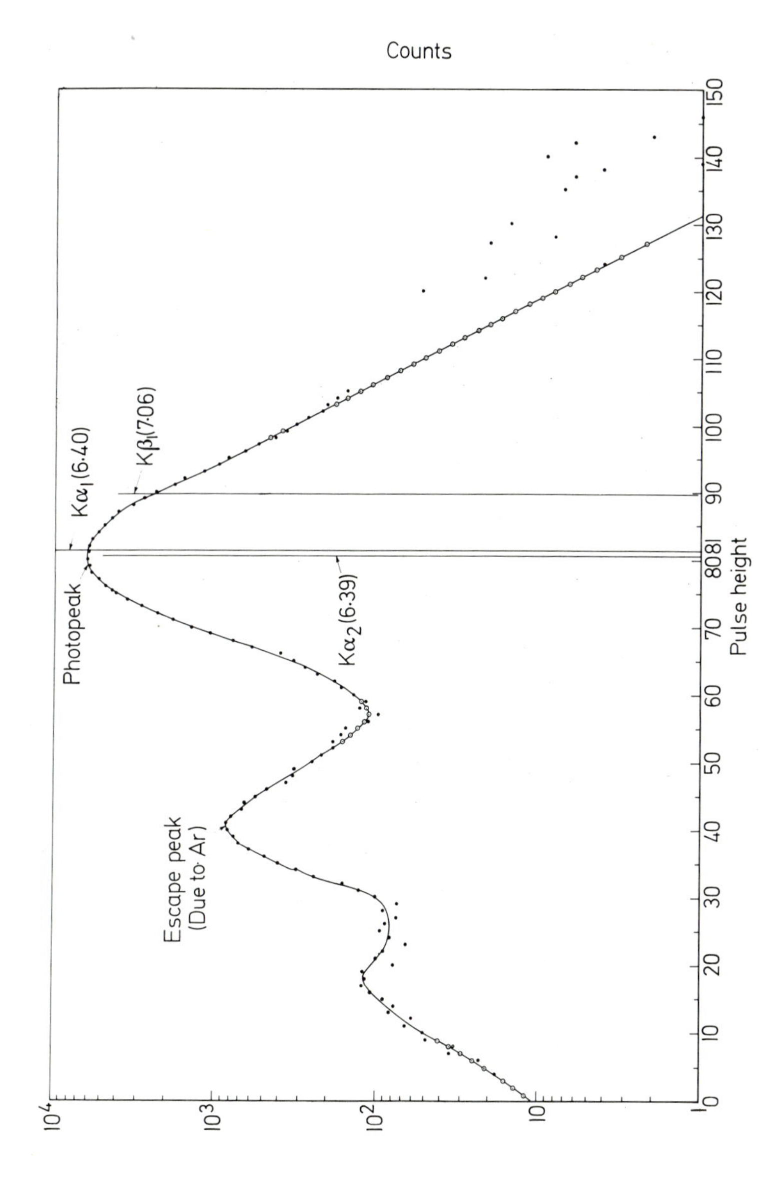

Figure 5.2 Discrete energy spectrum and corresponding pulse spectrum of the characteristic Fe X-ray lines. Pulse height spectrum measured with a sealed 90% argon 10% methane proportional counter

The pulse height spectrum obtained when monoenergetic gamma rays are detected using a scintillation system is never a line. Its shape is determined by gamma ray energy and source detector configuration. The shapes of these monoenergetic pulse height spectra are primarily determined by:

1. The relative magnitude of the photoelectric absorption, Compton scattering, and pair production cross sections (Bell, 1955) and,

2. the losses and statistical fluctuations that characterize the crystal, light collection, and photomultiplier system (Bell, 1955).

Let us examine the case where photoelectric absorption predominates, and Compton scattering and pair production are thought to be negligible.

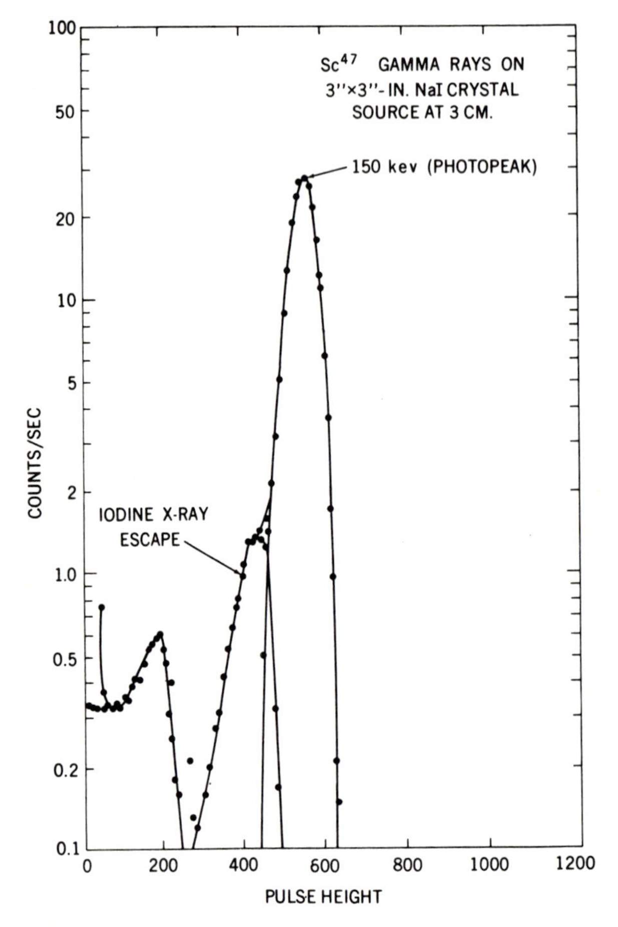

Figure 5.3 ^{47}Sc gamma rays on 3″ × 3″ NaI(Tl) crystal. Source at 3 cm

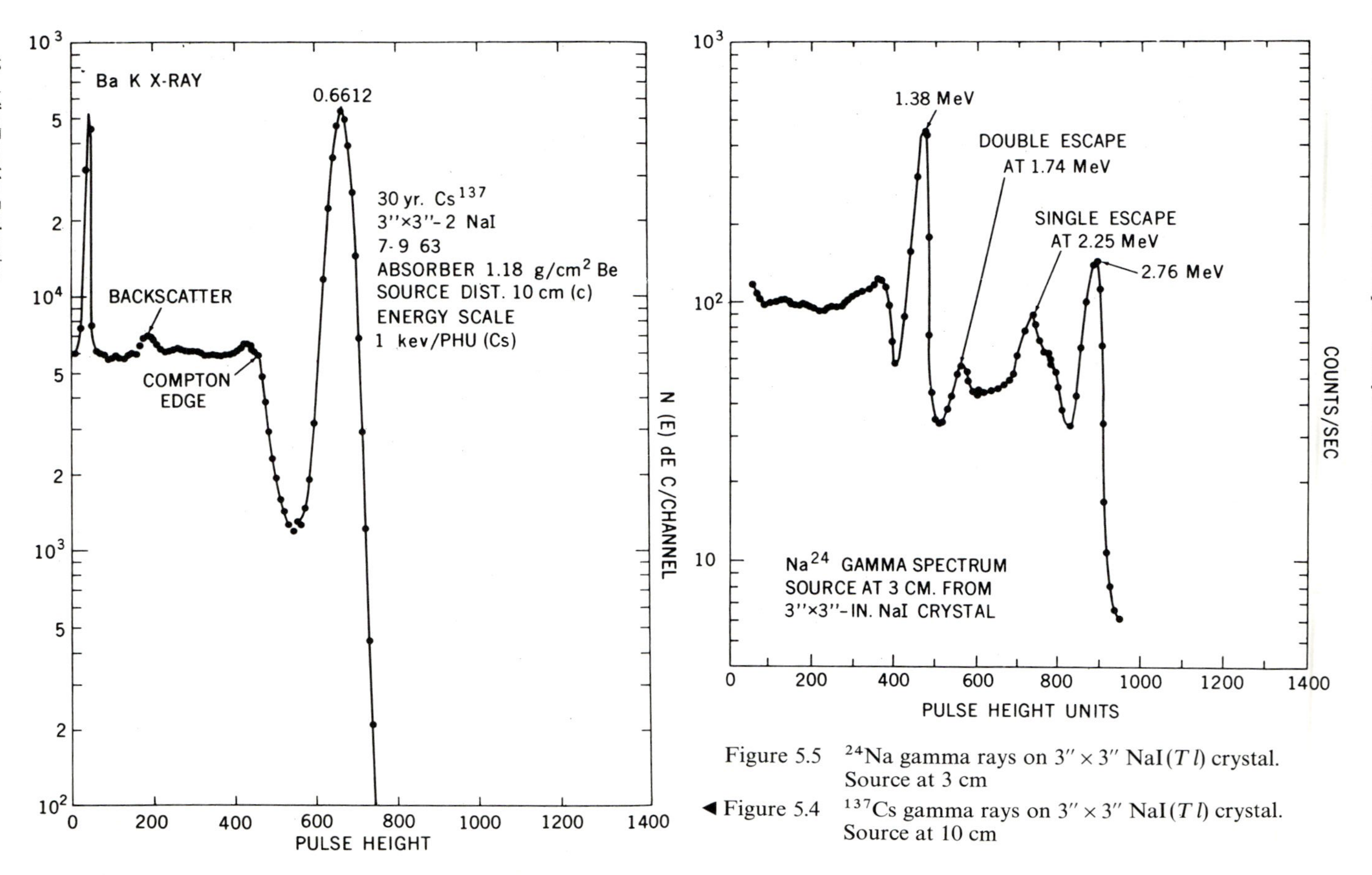

Figure 5.5 ^{24}Na gamma rays on 3″ × 3″ NaI(Tl) crystal. Source at 3 cm

◄ Figure 5.4 ^{137}Cs gamma rays on 3″ × 3″ NaI(Tl) crystal. Source at 10 cm

In this process, the kinetic energy imparted to a secondary electron is equal to the energy of the gamma ray minus the electron binding energy. This binding energy can be reclaimed in terms of the scintillation process by the absorption of the X-rays produced after photoelectric absorption. There is also the possibility that the X-rays may escape the crystal without being absorbed. The pulse height distribution caused by photoelectric absorption is characterized by two regions: the region of total absorption (the photopeak), and the region of total absorption minus X-ray escape energy (the escape peak). This distribution spreading, plus the Gaussian spreading previously discussed, yields a pulse height spectrum similar to that shown in Figure 5.3.

When Compton scattering becomes an important energy loss mechanism, another region, the so-called Compton continuum, is observed in the pulse height spectrum. In terms of the scintillation process, all the energy lost in scattering will be given up to the electron as kinetic energy. The gamma ray may lose part of its energy to the crystal; furthermore, after suffering a Compton collision or a number of Compton collisions, it may suffer a photoelectric absorption and lose its remaining energy. Thus the gamma ray either loses all of its energy in the crystal, or loses part of its energy in the crystal while the remainder of the ray escapes the crystal at a diminished energy (see Figure 5.4).

At energies higher than 2 MeV, pair production becomes appreciable. Two false "photopeaks" are then observed. Figure 5.5 is the pulse height spectrum of ^{24}Na. The gamma ray energies emitted by ^{24}Na are 2.76 MeV and 1.38 MeV. The three peaks of greatest pulse height are caused, in order of increasing pulse height, by:

1. Pair production with escape of both annihilation quanta.
2. Pair production with the absorption of one annihilation quantum.
3. Pair production with absorption of both annihilation quanta, and total absorption by photoelectric effect or any combination of other effects leading to total absorption.

In addition to the photopeak, the iodine X-ray escape peak, the Compton continuum, and the pair escape peaks, there are a number of other regions characteristic of experimentally determined monoenergetic pulse height spectra. These are:

1. The multiple Compton scattering region: Because of such scattering from materials surrounding the source and crystal, thus degrading the primary energy, there is a continuous distribution of gamma rays incident upon the crystal with energies less than the maximum energy. This tends to spread out the true Compton continuum produced by gamma rays of undegraded energy scattering in the crystal.

2. Annihilation radiation from the surroundings: Positrons emitted from the source may annihilate the surrounding material. Some of the

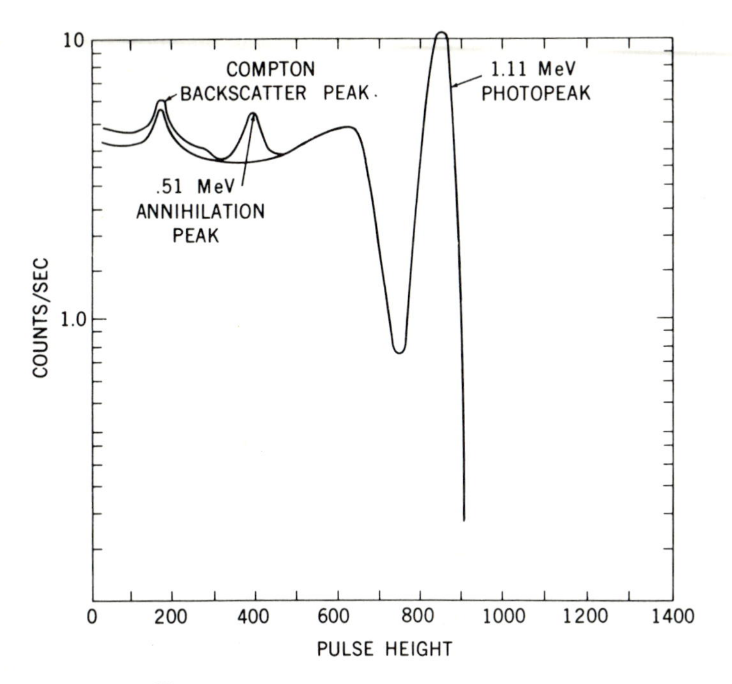

Figure 5.6 ^{65}Zn at 0.24 cm from NaI(Tl) crystal. Source at 10 cm

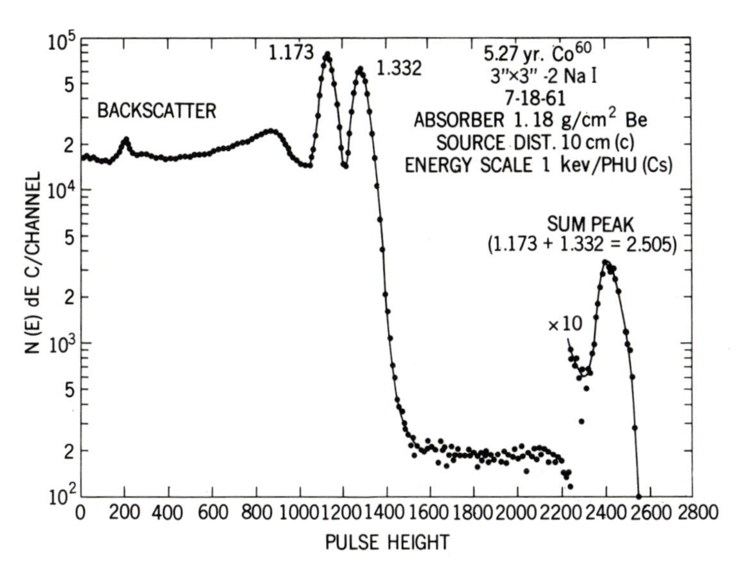

Figure 5.7 ^{60}Co gamma rays on 3″ × 3″ NaI(Tl) crystal. Source at 10 cm

0.51 MeV gamma rays produced in such a manner will reach the crystal, and a pulse height spectrum characteristic of 0.51 MeV gamma rays will be superimposed on the monoenergetic pulse height spectrum (see Figure 5.6).

3. Coincidence distribution: If two gamma rays interact with the crystal during a time which is shorter than the decay time of the light produced in the scintillation process, a pulse will appear whose height is proportional to the sum of energies lost to the crystal by both interacting gamma rays (see Figure 5.7).

Because the interaction time of both single and multiple interactions is shorter than the decay time of the light in the crystal, a single gamma ray interacting with the crystal produces only one pulse. The magnitude of the pulse height is affected by the type or number of interactions for a given gamma ray. Thus, if the above-mentioned coincidence effects are negligible, the measured monoenergetic pulse height spectrum can be considered as a distribution of the probability of energy loss as a function of energy for the given gamma ray energy and geometrical configuration. In addition, the shape of the monoenergetic pulse height distribution depends on the source detector geometry.

A semi-empirical method for calculating these shape functions for NaI (*Tl*) detectors has been developed and described by Heath et al. (1965).

Let us now develop a numerical method for obtaining the solution, $T(E)$, from a measurement of $Y(V)$ and a knowledge of $S(E, V)$, where $T(E)$ and $Y(V)$, respectively, are the differential energy spectrum and measured pulse height as previously described. Two cases will be considered: 1. where $T(E)$ can be described as a distribution of discrete energies; and 2. where $T(E)$ is a continuous function of energy.

For the first case, we write equation (1) as follows:

$$Y_i = \sum_{j=0}^{n} T_j S_{ij}, \tag{12}$$

where Y_i are the total counts in channel, i, of the measured pulse height spectrum, T_j are the total number of events occurring in an energy region Δj about j for the differential energy spectrum, and S_{ij} is a numerical description of the interaction function $S(E, V)$ and reflects the probability that if an energy, j, interacts with the detector it will appear as a pulse height, V, in the measured spectrum. Equation (1) reduces to equation (12) because $T(E)$ can be described as a discrete set of n energies with varying intensities, T_j. Both S_{ij} and Y_j are determined using an analyzer which samples both types of spectra mentioned above.

The Shannon Sampling Theorem discussed earlier is also applicable to the analysis of continuous distributions. This theorem, as we have indicated, states that a continuous distribution can be most completely described by a discrete set of values if the samples are separated by more than $\frac{1}{2} f_{max}$ (where f_{max} is the frequency of the most rapidly changing portion of the measured distribution). Because of the finite resolution of any measuring system, the maximum frequency will be set by the resolution of the given detector. If we again assume that the resolution function can be described by a Gaussian distribution; then, as previously shown, the criteria are reduced, requiring that the energy or pulse height increments needed to construct the function, $T(E)$, be chosen so that they are no greater than the half width at half maximum of the resolution function (detector response to a monochromatic energy). Thus, the distribution, $T(E)$, can be written as a set of numbers, T_j; the total number of events in Δj about j and Δj is equal to, or less than half, the resolution width at half maximum of the given detector for a given energy, j.

An example of how one chooses energy increments in practice follows: We shall assume the use of a scintillation detector which has a 10% resolution for 0.661 MeV gamma rays. The functional dependence of resolution on energy is given by equation (7), where the exponent, $n, = \frac{1}{3}$. Our interest is in a continuous spectrum ranging in energy from 0.073 MeV to 0.661 MeV. We begin by calculating Δj for $j = 0.661$ and for $j = 0.073$.

From the definition of resolution, R_E, as given in equation (6), we can write our sampling criterion for Δj as:

$$\Delta j = \frac{W_{1/2}}{2} = \frac{R_E \cdot E}{2} . \tag{13}$$

For $j = 0.661$ MeV, $\Delta_j = (0.10)\ (0.661)/2 = 0.033$ MeV. For the case where $j = 0.073$, we first calculate the resolution using the energy dependence in equation (7),

$$\frac{R\,0.073}{R\,0.661} = \left(\frac{0.661}{0.073}\right)^{1/3} = 1.44 .$$

Thus, the resolution at 0.073 will be 14.4%; at

$$j = 0.073, \quad \Delta_j = \frac{(0.144)(0.073)}{2} = 0.0053 \text{ MeV} .$$

Now from equation (13), and the energy dependence of R, we can write:

$$\Delta_j = (\alpha E^{-1/3}) | E = \alpha E^{2/3} . \tag{13a}$$

If we take the log of Δ_j we get:

$$\log \Delta_j = \frac{2}{3} \log E + \log \frac{\alpha}{2} . \tag{13b}$$

This is the equation of a straight line with a slope of $\frac{2}{3}$, when plotted on a Log Scale or full logarithmic scale. If one plots the two points just calculated in this manner (see Figure 5.8), one can then, from the straight line determined, obtain a discrete set for j_n which will be the minimum set required to describe the function, $T(E)$ (in this instance, T_j). Starting at $j_1 = 0.661$, we get $j_2 = 0.661 - 0.033 = 0.628$. In a similar manner, Δj_2 can be found in order to obtain J_3, etc. In general, $j_{n+1} = j_n - \Delta j_{n+1}$, and Δj_{n+1} can be determined graphically in the same manner until the energy range of interest has been covered. A more detailed discussion of the above method for use in gamma ray spectroscopy can be found in two papers by Trombka (1961, 1969).

Both discrete and continuous distribution can be described by an equation of the form shown as equation (12). We shall now solve equation (12) for T_j and obtain a general solution applicable to many of the techniques described here.

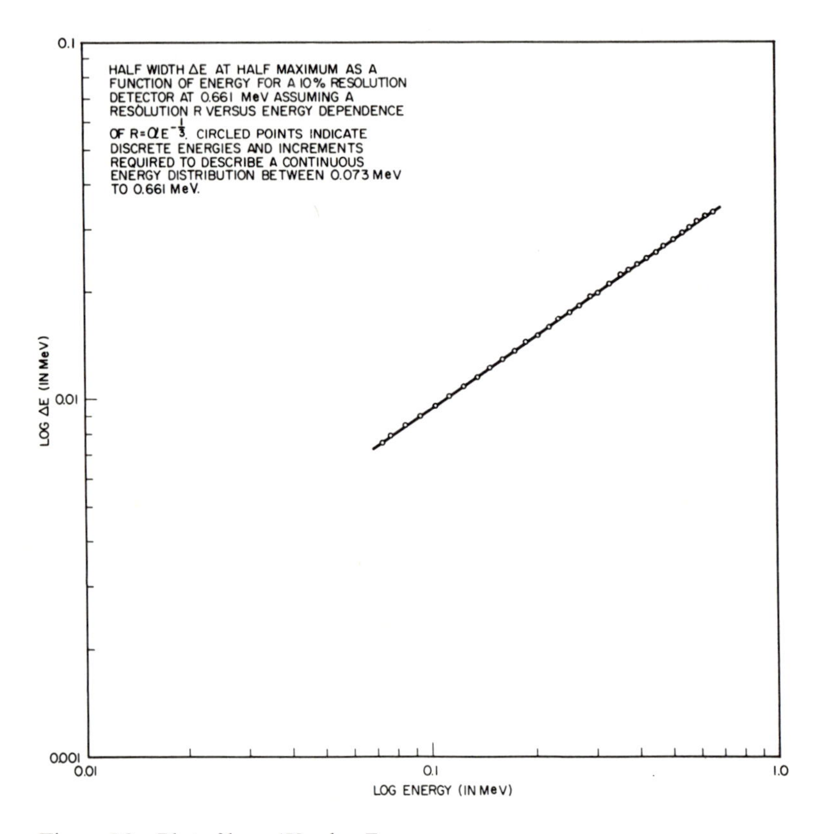

Figure 5.8 Plot of log ΔV vs $\log E$

A simple matrix inversion cannot be used to solve equation (12), for two major reasons. First, no exact solution for (12) exists because of the statistical nature of the counting procedure used in determining Y_i and S_{ij}. Second, because of the requirements derived from information theory (i.e., at least two samples per energy component required), one always obtains an over-determined set of equations; thus, an infinite number of solutions exist. The problem, therefore, is to select one solution out of this infinite set which most reasonably corresponds to the real world.

Equation (14) represents the least-square criteria used to find the most probable value of T_j,

$$M = \sum_i \omega_i (Y_i - T_j S_{ij})^2 \rightarrow \text{minimum}, \tag{14}$$

where ω_i is the statistical weight for each channel, i, and $\omega_i \sim 1/\sigma_i^2$. In formulating the linear least-square method, it is assumed that the pulse height scale does not vary between S_{ij} and Y_i, or among the various components of S_{ij}; therefore, the minimum can be found by taking a partial derivative with respect to some component, T_k. The derivative is then set equal to zero,

$$\partial M / \partial T_k = 2 \sum_i \omega_i (Y_i - \sum T_j S_{ij}) S_{ik} = 0\,. \tag{15}$$

This can be written in matrix form (Trombka and Rose (1953) as:

$$\tilde{S} \omega Y - (\tilde{S} \omega S) T = 0\,, \tag{16}$$

or, solving for T,

$$T = (\tilde{S} \omega S)^{-1} \tilde{S} \omega Y, \tag{16a}$$

where S is an $m \times n$ matrix of the possible components present, $\tilde{S}$ is the transpose of S, ω is a diagonal matrix of the weighting functions, and Y is a vector describing the measured spectrum.

The remainder of the discussion will be devoted to finding the proper algorithm for solving equation (16a). The algorithm to be used will strongly depend on the particular type of spectrum to be analyzed (e.g. gamma ray, X-ray, mass spectroscopy, etc.). Certain general considerations will be presented.

Calculation of Errors

Once the T's in equation (16a) have been determined, it is possible to determine the mean-square deviation in T. If it is assumed that the $S_{i\gamma}$ (i.e., the magnitude in the normalized standard spectrum of the component γ in channel i) is without error, the error in the T_γ calculation can then be obtained under the assumption that the statistical variations in the determination of the T_λ's are due to the statistical variation in the measurement of the Y_i. Equation (16a) can also be written:

(16b) $$T_\lambda = \sum_i \sum_v C_{i\lambda}^{-1} S_{iv} \omega_i Y_i ,$$

where $C=(\tilde{S}\omega S)$ and is a symmetric matrix; the elements of C are given by

(17) $$C_{v\gamma} = \sum_i \omega_i S_{iv} S_{i\gamma} .$$

From equation (16a), it is seen that T_λ is a linear homogeneous function of the counts, Y_i. Thus the mean-square deviation, $\sigma^2(T\lambda)$ corresponding to the variation in Y_i can be written as:

(18) $$\sigma^2(T_\lambda) = \sum_i \sum_v \sum_\gamma C_{\gamma\lambda}^{-1} S_{iv} S_{i\gamma} \omega_i^2 \sigma^2(Y_i) .$$

We now can consider the nature of $\omega_i = \dfrac{b}{\sigma^2(Y_i)}$. Equation (18) reduces to

$$\sigma^2(T_\lambda) = b\, C_{\lambda\lambda}^{-1} ;$$

that is, $\sigma^2(T_\lambda)$ can be found from the diagonal elements of the C^{-1} matrix. The goodness of fit can also be calculated and is used throughout the calculation. This is the so-called chi-square value and is shown to be (Bennett and Franklein, 1954).

(19) $$\chi^2 = \sum \omega_i \frac{(Y_i - \sum_\lambda T_\lambda S_{i\lambda})^2}{n-m} ,$$

where n is the number of channels, and m is the number of components used in the fit. Thus $n-m$ is the number of degrees of freedom. It can be shown that the error calculated in (T_i) is true only if chi-square $=1$. If not, then

(20) $$\sigma^2(T_i) = b\chi^2 C_{\lambda\lambda}^{-1} .$$

This is the variance used in the calculations to follow.

It can be shown further that the off-diagonal elements, $C_{\gamma\lambda}^{-1}$, are the covariance between the γth and λ^{th} component (Scheffe, 1959; Bennett and Franklein, 1954); the percentage of interference, $F_{\gamma\lambda}$, can then be calculated in the following manner:

(21) $$F_{\gamma\lambda} = \left\{ \frac{(C_{\gamma\lambda}^{-1})^2}{(C_{\gamma\gamma}^{-1} C_{\lambda\lambda}^{-1})} \right\} \times 100\,\% .$$

The $F_{\gamma\lambda}$ can also be considered as a measure of whether the set of library components, S, is truly an orthogonal set.

Method of Solution Using Non-Negativity Constraint

Solving equation (16a) directly by simple matrix multiplication and inversions sometimes results in negative solutions for the T_γ's. These solutions are often real in terms of the definition of the least-square, and are acceptable. However, what happens if the negative solutions are caused by problems in the inversion of the matrix, $(\tilde{S}\omega S)$? This can be attributed, for example, to the fact that the library function cannot be determined to a sufficient number of significant figures compatible with the accuracy required in obtaining $(\tilde{S}\omega S)^{-1}$. Negative solutions may then cause oscillation. These oscillations can be damped out by using a non-negativity constraint (i.e., setting all components that go negative to zero before oscillations can occur). One comment before proceeding, the non-negativity constraint must be utilized with great caution. For example in those cases where the composition is known *a priori*, the known elements are used as the library components and negative solutions occur for the most part because of correlation effects and/or because the variance on the relative intensity in the pulse height spectra due to that negative component is greater than the magnitude of the relative intensity determined by the least-square method.

From past experience in the utilization of this technique it has been found that the non-negativity constraint is extremely useful in those cases where the library contains more components than are present in the mixture (i.e., the composition is not known *a priori*). The task, then, is to find the subset of the library components which will best fit the unknown pulse height spectrum.

A method for introducing constraints into the solution of such equations was discussed by Beale (1959) and Trombka (1963), and is equivalent to the following matrix approach. Assuming that the measured distributions is made up of only two components (e.g., the S_{i1}'s and the S_{i2}'s), one can use least-square fitting to obtain the T_1 and T_2 from equation (16b). It has been assumed that $T_3 = T_4 = \cdots = T_n = 0$. If $T_1 > 0$ and $T_2 > 0$, then one adds a third component and solves for T_1, T_2, and T_3. If any of these T's are negative, they are set equal to zero and the corresponding library components are eliminated from the library matrix. In this way, each component is tested until only a positive or zero magnitude is found for all the components under consideration. Each least-square fit (i.e., as each new component is tested) is made to the total pulse-height spectrum. Because the number of components used in a given least-square fit is less than, or equal to, the actual number of components in the mixture, the relative intensities (the T's) obtained will be greater than, or equal to, their true values. As the components are added to the library, the magnitudes of the T's will decrease and approach their true value. Thus if a component when added to the library yields a negative magnitude for T, the addition of more components will make this value decrease or become more negative. In this way the solution under the non-negativity constraint is obtained.

Compensation for Gain-Shift and Zero-Drift

Gain-shift can be considered as a compression or expansion of one pulse height scale with respect to the second scale, whereas zero-drift can be considered as a linear displacement of one scale with respect to the other. Thus, in general, these two changes can be expressed as

$$P_{ji} = g\,P_{ki} + \varepsilon, \tag{22}$$

where P_{ji} is the channel number, i, for spectrum j, P_{ki} is the channel number, i, for spectrum k, g is the gain-shift, and ε is the zero displacement. In order to perform the linear least-square analysis, g must equal 1, and ε must equal zero.

The spectrum to be changed is described by P_{ki}, the total counts in channel P_{ki} of spectrum k. To change the kth spectrum so that it is on the same scale as the j spectrum, the following procedure is used (if g and ε are known). The pulse height scale, P_{ki}, is multiplied by g, and the intensity scale, P_{ki}, is divided by g. It must be remembered that the pulse height spectra can be considered as histograms. The integral under the pulse height distribution must remain constant. The procedure for compensating for gain shift will keep this area a constant.

The spectra used in the least-square analysis are included for integer values of pulse height which are separated by $\Delta P_{ji} = 1$. The gain is shifted by some value g, which could produce intensity, Y_{ji} values at fractional values of pulse height, P_{ki}/g, and cause ΔP_{ki} to become either less than, or greater than, unity. Linear extrapolations between adjacent points in the pulse-height distribution are used to find the intercepts for the integer pulse-height values.

Compensation for a zero drift is accomplished by changing the pulse height scale, $g\,P_{ki}$, to $g\,P_{ki} + \varepsilon$. There need be no operation on the intensity values, Y_{ki}, for this linear displacement does not change the value of integral under the pulse height distribution. Compensation for Y_{ki} being at noninteger values of $g\,P_{ki} + \varepsilon$, and for the fact that $\Delta(g\,P_{ki} + \varepsilon)$ may not be unity, is made by using the linear extrapolation method previously mentioned. Figure 5.9 shows the results of gain shifting on a spectrum with $g \approx 0.70$ to match the library spectrum. The spectrum was obtained using a proportional counter filled with P-10 gas*, and 256-channel

* A 90% argon and 10% methane mixture.

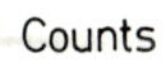

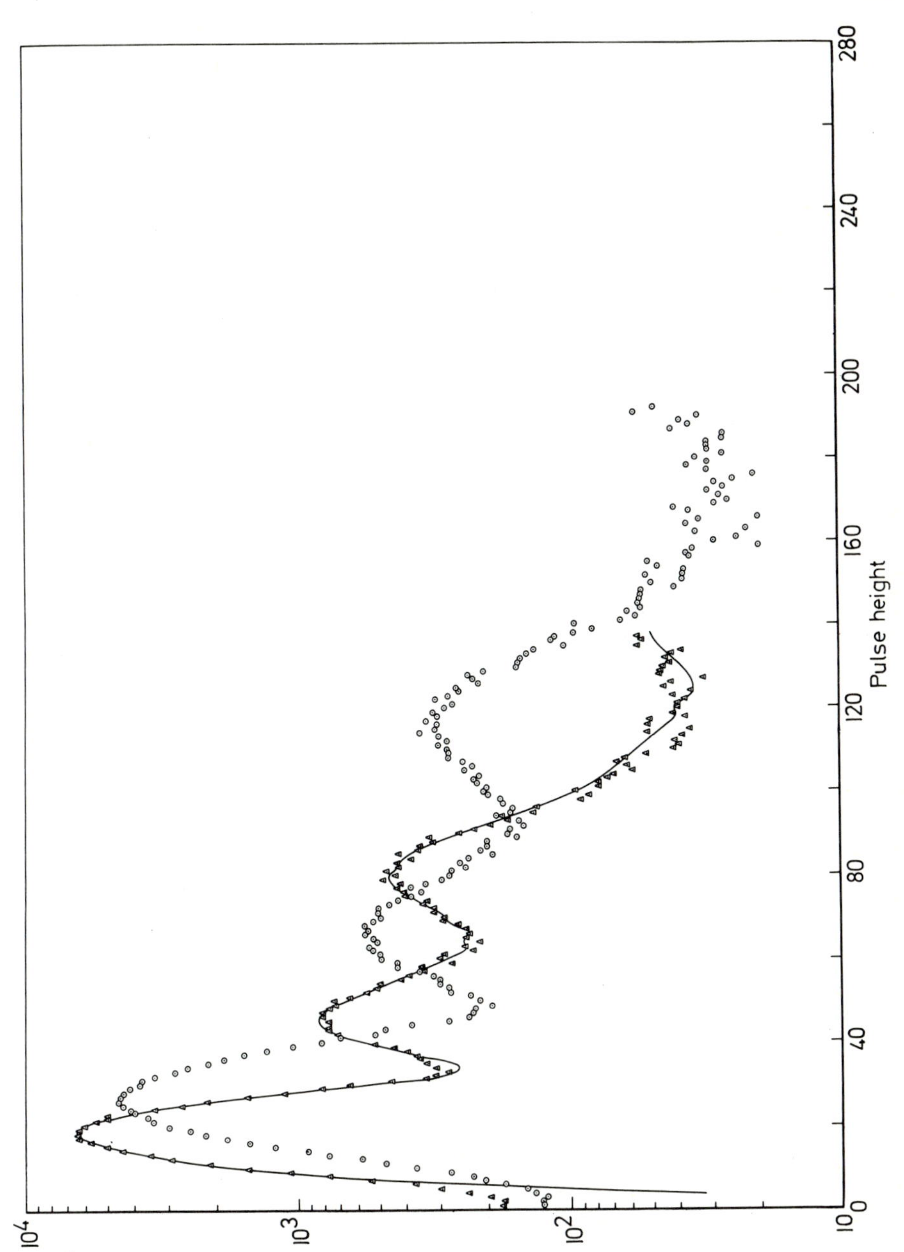

Figure 5.9 Effect of gain shift; sample of silty sand from Hopi Butte, Arizona

analyzer. Characteristic X-rays produced by α particle bombardment of sand samples were measured. The dots represent the measured spectrum; the triangles represent the gain shifted spectrum; the solid line shows the synthesized spectrum using the least-square fit. The agreement is extremely good.

The computer program developed at this laboratory* will compensate for a constant value of gain-shift or a zero-drift, or, given a lower and upper limit for a gain-shift or zero-drift, will search for the proper value between the limits given. The criterion used to find the best gain-shift, g, and zero-drift, ε, in the given interval is that values g and ε be chosen such that chi-square be defined in equation (19) as a minimum. The procedure used to find this minimum chi-square is as follows. The least-square fit is performed for the minimum and maximum values of g in the given range. Then a fit is made for a value of g in the middle of the range. Again fits are made for values of g in the interval midway between the upper and middle value of g, and then the middle and lower value of g. The chi-square obtained each time as always compared with the smallest value obtained previously. This process of halving the range and determining the smallest chi-square for these values of g is continued until the difference between two successive tests is smaller than some predetermined limit. In the case of the program developed at this laboratory, experience has shown that this number is 0.01. Once the value of g has been found, this gain-shifted spectrum is used as the input spectrum to the zero-drift search program. The same iterative process is used to determine the best value of ε in the zero-drift program as was used for the gain-shift search.

The precision to which the values can be determined will depend strongly on the resolution of the detector system because the better the resolution, the sharper and steeper will be the shape of the library function and the measured spectrum. The sharper the shape of these functions, the greater will be the change in chi-square for small changes in g or ε. Furthermore, it has been found that the gain-shift has greater sensitivity for the higher pulse height region while, for the zero-drift procedure, the lower pulse height part of the spectrum shows greater sensitivity.

Preparation of the Library

To obtain reasonable results with the technique described in this paper, the library functions (i.e., S_{ij}'s used in equation (16)) must be known as well as possible. Careful preparation of the library standards, long

* Copies of the program may be purchased from COSMIC (Computer Software Management Information Center) University of Greorgia, Athens, Ga., USA.

counting times and repeated measurement of these standards, and proper background and linearity compensation are necessary to obtain a set of library functions which will yield meaningful results. The error in the knowledge of the library function must be significantly less than the error in the measurement of the complex spectrum to be analyzed.

When we talk of the linearity problem, we mean that the energy pulse height scale must be the same for all of the library functions used in a given analysis. Even when great care is taken in measuring the standards, small gain-shifts and zero-drifts occur so that there may be small changes in the scale. Two approaches may be used to make the correction. First of all, the energy versus pulse height scale can be determined carefully, and then each component can be checked with respect to this scale, and, using gain-shift and zero-drift programs again, each component corrected

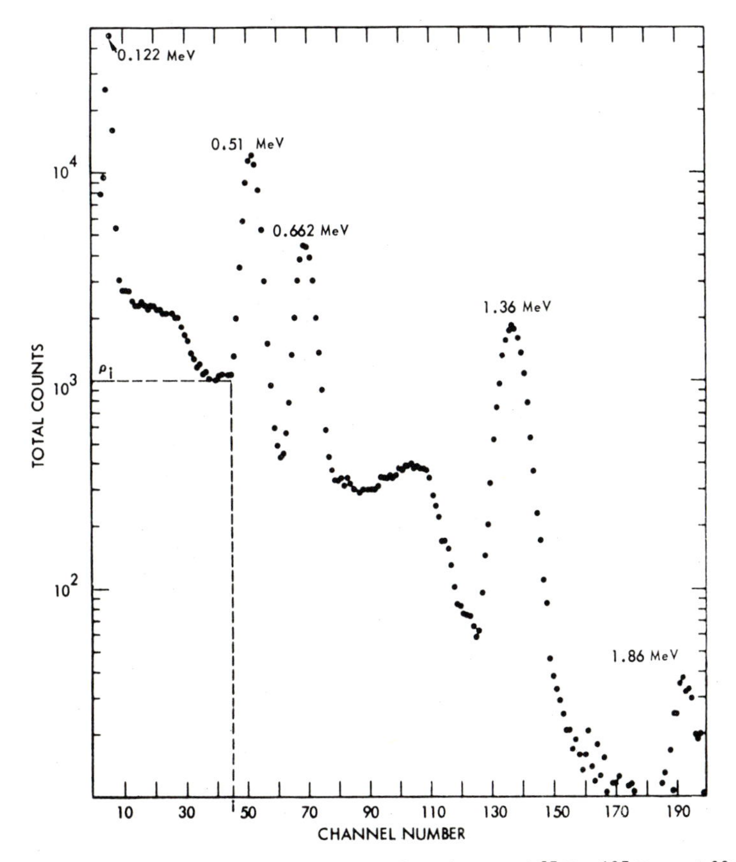

Figure 5.10 Pulse height spectrum of a mixture of ^{57}Co, ^{137}Cs, and ^{22}Na 10 cm from a 3″ × 3″ NaI(Tl) crystal

for a common scale. The second approach is the one used in this method. Each monoenergetic or monoelemental spectrum and a number of mixtures of these elements are measured separately. The monoelemental components are then varied (both gain-shifted and zero-drifted) separately in order to obtain a best fit to the mixture spectrum.

The adjusted spectra can then be used in the main program, with quite an improvement. Table 5.1 shows the results of the least-square fit before and after library adjustment. The spectrum analyzed is the mixture spectrum shown in Figure 5.10. The gain-shift and zero-drift for each component, and the true relative intensity of the mixture, are also indicated. The great improvement in the chi-square value indicates the greater confidence in the solution. Furthermore, with even these small gain-shift and zero-drift compensations, there is a great improvement.

Table 5.1. *Results of the least-square analysis of the mixture of* ^{57}Co, ^{137}Cs *and* ^{22}Na *before and after the application of the library preparation technique*

Monoelemental component	Gain-shift	Zero-drift	True relative intensity	Relative intensity before library adjustment	Relative intensity after library adjustment
^{57}Co	1.002	−0.004	0.200	0.185±0.013	0.193 ±0.005
^{137}Cs	1.002	−0.074	0.200	0.099±0.013	0.202 ±0.002
^{22}Na	1.048	0	0.100	0.102±0.004	0.1001±0.006
chi-square				114	2.2

Correlation and Resolution

The number obtained by using equation (21) is a measure of how differently, or how well, one component or library spectrum can be resolved with respect to a second. It can also be considered as a measure of whether the components in the library of standard spectra can be considered orthogonal. To illustrate how equation (21) is used, assume that the library function or monoelemental function can be described by Gaussians:

$$S_{i\gamma} = \exp\left\{\frac{-\frac{1}{2}(i-P_\lambda)^2}{\alpha_\lambda^2}\right\}, \tag{23}$$

where $S_{i\lambda}$ are the counts in channel i due to the λth component, α_λ^2 is a measure of the width of the Gaussian for energy, λ, and $P\lambda$ is the position on the pulse height spectrum corresponding to energy, λ.

Now let us calculate the percentage of interference between two monoenergetic pulse height spectra with Gaussian form using equation

(21) and a shape given by equation (23). If we assume that $\omega=1.0$, equation (19) becomes

$$(24) \qquad F_{v\lambda} = \frac{\left\{\sum_i (S_{i\gamma} S_{i\lambda})^2\right\}}{\left\{\sum_i (S_{i\gamma}^2) \sum_i (S_{i\lambda}^2)\right\}}.$$

Using equations of the form given in equation (23), and replacing the summation by an integral from $-\infty$ to $+\infty$, equation (24) becomes:

$$(25) \qquad F_{\gamma\lambda} = \left\{\frac{2\alpha_\lambda \alpha_\gamma}{\alpha_\lambda^2 - \alpha_\gamma^2}\right\} \exp\left\{\frac{-(P_\lambda - P_\gamma)}{\alpha_\gamma^2 - \alpha_\lambda^2}\right\}.$$

Defining the resolution, R_λ, as before,

$$(26) \qquad R_\lambda = \frac{W_{\frac{1}{2}}}{P_\lambda}.$$

Furthermore, assuming that the resolution, R_λ, is inversely proportional to some power of energy or pulse height, P_λ; that is:

$$(27) \qquad R_\lambda \approx P_\lambda^{-n},$$

and if it is assumed that $n=0.5$, then

$$(28) \qquad \frac{R_\lambda}{R_\gamma} = \left(\frac{P_\gamma}{P_\lambda}\right)^{\frac{1}{2}}.$$

Furthermore α_λ can be rewritten in terms of R_λ and P_λ as $S_{i\lambda}=\frac{1}{2}$.
Then,

$$(29) \qquad \alpha_\gamma = \frac{\frac{1}{2} R_\gamma P_\lambda}{(2\log 2)^{\frac{1}{2}}}.$$

Substituting the relationships in equations (28) and (29) into equation (29), we get for this special case:

$$(30) \qquad F_{\gamma\lambda} = \left\{\frac{2(P_\lambda P_\gamma)^{\frac{1}{2}}}{P_\lambda + P_\gamma}\right\} \times \exp\left[\frac{-P_\gamma (P_\lambda - P_\gamma)^2}{\{(P_\lambda - P_\gamma)\alpha_\lambda^2\}}\right].$$

Figure 5.11 is a plot of equation (30) for various detector resolutions as a function of the percentage of separation $(P_\gamma - P_\lambda)/P_\lambda$ for decreasing P_γ. If $P_\gamma = P_\lambda$, the Gaussians are identical, and the percentage of interference is 100%. As the percentage of separation increases, the percentage of interference decreases. The number, $F_{\gamma\lambda}$, is (in this case) thus a measure of how well two Gaussians can be resolved.

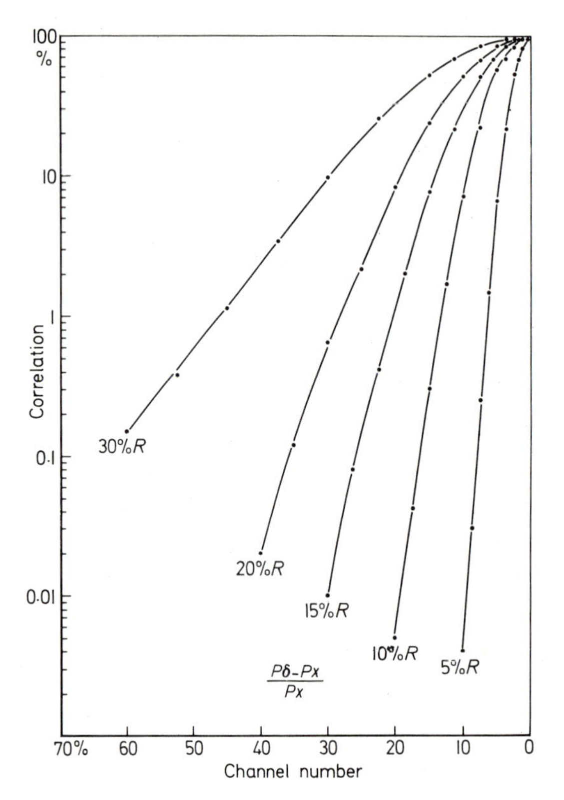

Figure 5.11 Percentage of interference for various resolutions

If two functions do interfere strongly, it is sometimes possible to impose a physical constraint which can eliminate the interference. For instance, in activation analysis utilizing gamma-ray spectroscopy, a physical constraint which depends on the half-life of the various nuclear species in the mixture being analyzed can be used. The following procedure is used for the case of two nuclear species with different half-lives. Assume that the half-life of the second of these two components is the longest. Pulse height spectra are obtained as a function of time. Spectra are measured until the short half-life has decayed out, for all practical purposes. A number of measurements are made for the longer half-life component. When the short half-life component has decayed out, the correlation between the two components will go to zero. The intensity of the second component and its corresponding statistical variance can now be determined. Knowing the relative intensity as a function of time and its half-life, an estimated intensity of the second component can be determined for the earlier times. The second component can then be subtracted out for those earlier times when both components were present; the analysis is repeated with the second component eliminated from the library, and the relative intensity of the first component is determined with the correlation eliminated. Trombka and Adler provide an example for the case of X-ray analysis in another work (1967). This constraint must be employed with great caution. Because the extrapolated values used in the subtraction are only estimated values, some residual amount of the component being subtracted may still be present in the resultant spectrum.

Thus the estimated error of the extrapolated value, and possibly the effect of the component correlation, should be used in the least-squares analysis of the spectrum after subtraction. In the techniques described herein, the estimated error of the extrapolated value is used in the calculation of the weighting factors (the "omega" matrix) and therefore included in the determination of both the relative intensities and their variance obtained from least-square analysis of the subtracted spectrum.

Background Compensation

A number of problems tend to complicate background compensation. Examples are possible shift of the pulse height scale with respect to either the library function scale or the scale of the measured unknown spectrum, possible nonlinear effects which complicate the problem of finding the proper intensity of background to subtract, and the inclusion of the statistical error due to background subtraction in the calculation of the least-square fit.

The first problem, background gain-shifts and zero-drift, can be handled in a number of ways. In one method, the technique used for the library preparation, the pure background spectrum is measured and then a second mixture spectrum is measured. This mixture spectrum is obtained using a source which is characteristic of one of the standards in the library. This second standard spectrum must be low enough in energy so that there will be no significant correlation between these two functions (i.e., the standard and background), also, that these functions can be considered as linearly independent. The mixture spectrum is then measured, and the standard library function and the pure background spectrum are used and adjusted using the library preparation program so that a best fit to the mixture spectrum is obtained. In this way, the gain shift and zero-drift factors can be determined. Provision is made in the least-square analysis program to make such constant adjustment of the pulse height scale for background before subtraction from the measured complex spectrum to be analyzed.

Alternately, if it is assumed that the background has shifted with respect to the library exactly as has the complex spectrum to be analyzed, there is provision in the program to vary the background spectrum between the same limits as prescribed for the spectrum which is being analyzed; a search for a minimum chi-square helps to determine the best gain-shift and zero-drift values for both. These routines for gain-shift and zero-drift compensation before going into the main least-square analysis, and those later steps to be described, can be considered as data input preparations for the main program.

Now consider the nonlinearity in determining the intensity of the background spectrum to be subtracted. This problem is encountered in the analysis of pulse height spectra obtained in nondispersive X-ray fluorescence analysis (Trombka and Adler, 1970). The background in this problem can be considered as comprised of two components: one attributed to the natural radioactivity in the surroundings, and a second caused by coherent scattering from the sample of X-radiation originating at the excitation source. It is the fluorescent X-radiation produced in the irradiated sample, not the scattered radiation, which is of interest in the analysis. The amount of background radiation due to scattering will be affected by the nature of the sample (average atomic number, density, etc.). Thus the background to be subtracted will be determined not only by the scattering source but also by the sample being irradiated.

The background spectrum is included as a library component and used in the least-square fit. In practice, this approach can be used only when a significant portion of the pulse height spectrum can be attributed to the background only. If this were not the case, the background spectrum would strongly correlate with many of the monoenergetic or monoelemental library components and a unique solution would not be obtained.

The final problem concerning the error due to background subtraction is now considered. In calculating the statistical weight, ω, and the variance in the relative intensity, T, the increase in statistical variance due to background subtraction is automatically included in the computer program. This correction is made only if the background subtraction is performed in the program. If, for example, the subtraction is done on the multichannel analyzer, and all that is available is the spectrum with background subtracted, then the calculation of the statistical error from the diagonal elements of inverse matrix, $(\tilde{S}\omega S)^{-1}$, will not be correct for ω. The program inputs the proper information on error calculations which are done outside the program. This is done essentially by reading in a corrected ω matrix. In the case where the background is included as a library element, proper compensation for statistical variance in the background will be difficult if not impossible. Therefore, in order to use this technique properly, the background must be measured so that the statistical variance in measurement of the background is significantly less than in the spectrum to be analyzed.

Application of Least-Square Techniques to Experimental Problems

Discrete Spectra

The solution of an experimental problem in activation analysis will now be presented. A set of standard library measurements was obtained by irradiating Na, Cl, K, Mn, Sc, Ar, As, Cu, Cr, I, and La in a nuclear reactor. Figure 5.12 gives an example of such a spectrum. Background spectra were also run. The library and background were adjusted to be on a common pulse height scale, and the contaminants indicated in the illustrations of library components were subtracted. A mixture of these elements was prepared and then irradiated in the reactor. Gamma-ray pulse height spectra of these mixtures were then measured from time

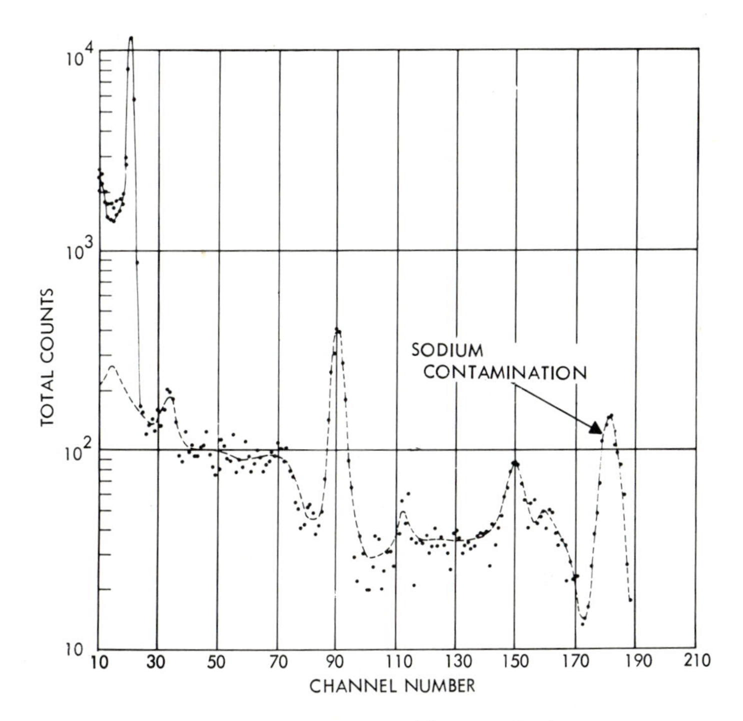

Figure 5.12 Pulse height spectrum of ^{51}Cr standard

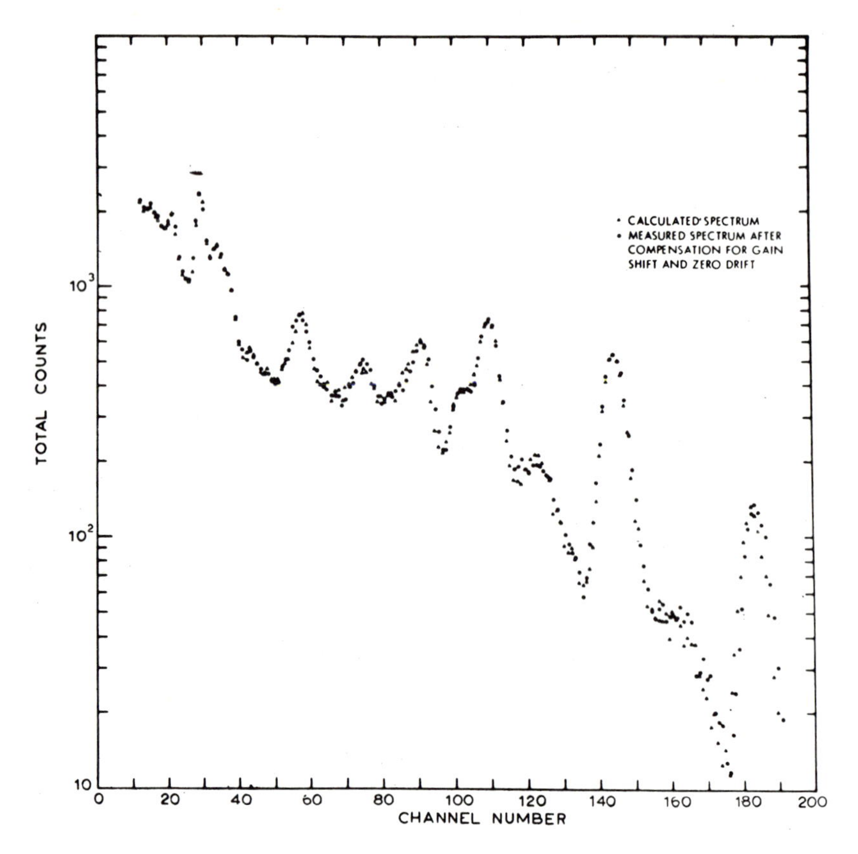

Figure 5.13 Pulse height spectrum of mixture no. 41122

periods of a few minutes to a number of days after exposure (Figure 5.13). The dots are the measured spectra, and the triangles are the synthesized spectra obtained from the least-square fit with gain-shift and zero-drift adjustment.

A tabulation of the monoelemental components that significantly correlated, using the least-square method with the non-negativity constraint, is shown with the percentage of interference obtained in Table 5.2. It was then decided to determine the intensities of the Na, Sc, and La components. From the half-life curves, the estimated relative intensities of these components were determined for the earlier time. These components were then subtracted from the nine pulse height spectra, and the relative intensities for Cl, Mn, K, Cu, and Cr were recalculated as a function of time. Using the non-negativity constraint, the argon component was rejected each time and set equal to zero.

Arsenic and iodine did not correlate significantly with any other component; therefore the values obtained before component subtraction were considered to be correct. An example of a half-life curve is shown in Figure 5.14. The curve was obtained using the results of the least-square analysis.

Table 5.2. *Interference*

Element	Na	Sc	La
Cl	14.5		
Mn		40.5	
K			21.4
Cu			13.4
Cr			12.9

The results of the calculation are shown in Table 5.3. The error shown is obtained from counting statistics. The errors due to changes in flux

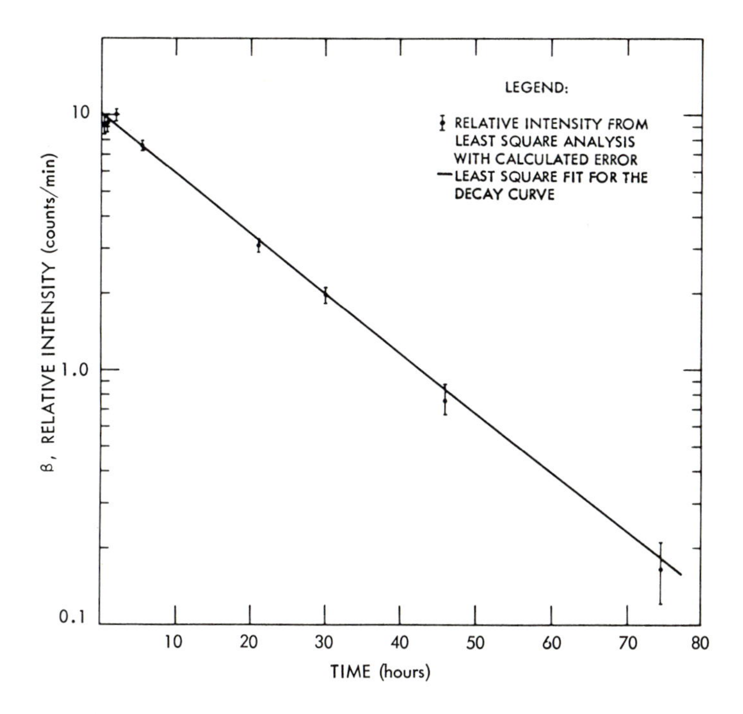

Figure 5.14 Decay curve for ^{42}K

and irradiation time are not included. Furthermore, the Cr standard sample was found to have Na contamination. Thus the prepared value had to be smaller than the calculated value; from the estimate of the amount of contamination present, it was found that the $7.8 \pm 1.5\ \mu g$ was closer to the true amount. The larger difference between the experimentally determined value and the prepared value may also be attributed to the fact that the standard spectrum obtained required the subtraction of the Na component. This increased the error in the determination of the relative intensity of the Cr component. This measured error is not included in the error calculation presented in Table 5.3. Furthermore, slight argon contamination had to be subtracted from the Mn, Sc, and I standard spectra. These errors plus the error in obtaining an inversion factor from relative intensity to μg or mg are not included in the error calculation presented in Table 5.3. With these factors in mind, the experimentally determined values seem to agree well with the values used in preparing the unknown sample.

Table 5.3. *Quantitative and qualitative analysis obtained with activation analyses*

Element	Prepared sample	Experimentally* determined by least-square analysis
Na	5.00 μg	7.8 ±1.5 μg**
Cl	36.8 μg	39.5 ±1.5 μg
K	0.404 mg	0.371 ±0.013 mg
Mn	0.101 μg	0.090 ±0.008 μg
Sc	14.2 μg	13.6 ±0.4 μg
As	5.00 μg	4.97 ±0.18 μg
Cu	2.23 μg	2.70 ±0.21 μg
Cr	1.82 mg	1.77 ±0.03 μg
I	2.49 μg	2.45 ±0.12 μg
La	2.29 μg	2.65 ±0.14 μg

* Errors shown are only those attributed to counting statistics.

** This value seems to be correct because of Na contamination of the Cr standard.

Continuous Spectra

As a demonstration of the method of analysis as applied to continuous spectra, the following experiment was performed. A pulse height spectrum of gamma rays caused by neutron radiative capture, neutron activation, and natural background was measured. The spectrum is shown as curve (a) in Figure 5.15. This spectrum was obtained using a 14 MeV generator (*D-T* reaction) to irradiate a large sample of basalt,

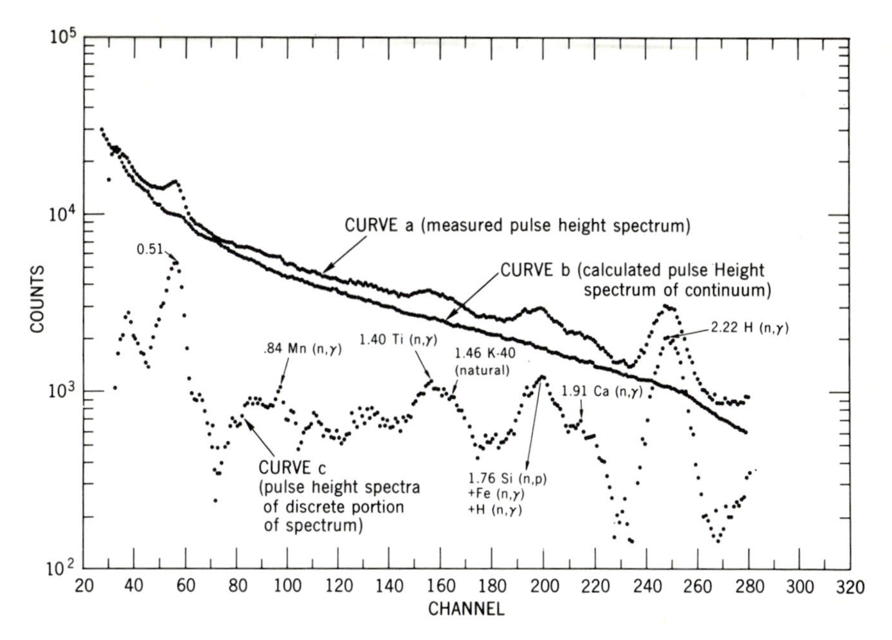

Figure 5.15 Pulse height spectra of basalt sample irradiated with pulsed accelerated source of 14 MeV neutrons using a 3″ × 3″ NaI(Tl) detector

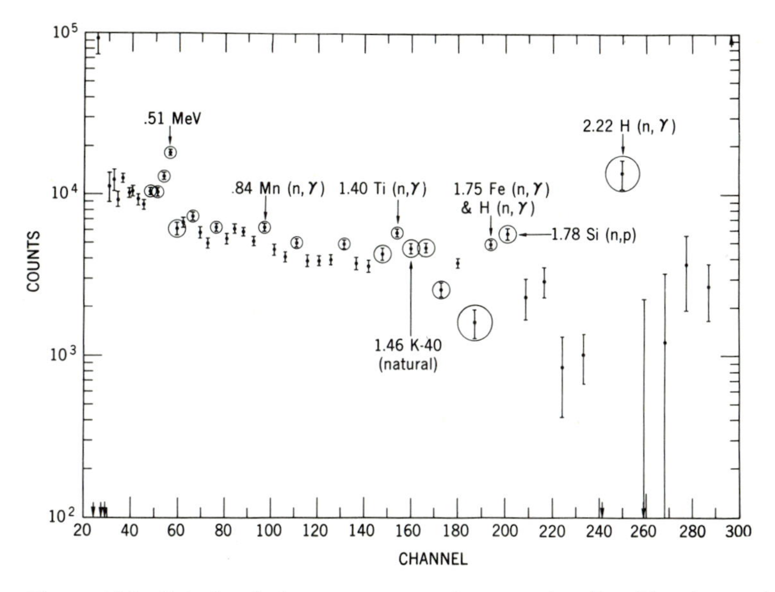

Figure 5.16 Calculated photon spectrum of neutron irradiated basalt sample

and a $3 \times 3''$ NaI (Tl) detector to measure the gamma rays in the energy region from 0.3 MeV to 2.5 MeV. Because the irradiated sample was rather large, many of the gamma rays produced in the basalt were scattered thus producing a large scattered continuum which tended to mask the discrete line spectrum.

The following is an outline of the procedure used to determine the discrete photon spectrum which is used, in turn, for the compositional analysis:

1. Equation (16a) is employed as a start in obtaining the differential energy (photon) spectrum. The matrix, S, contains response functions for energies from 0.300 MeV to 2.5 MeV at intervals determined from the Sampling Theorem considerations presented at the beginning of this chapter. The differential photon spectrum is shown in Figure 5.16.

2. The spectrum shown in Figure 5.16 still contains the effects of the scattered radiation. To eliminate these effects, a pulse height spectrum of the scattered radiation is derived:

a) Identify by observation all possible discrete lines (circled points in Figure 5.16). The scattered radiation is generally a slowly varying function of energy, thus any points statistically above or below adjacent points can be considered as a possible discrete line or an oscillation in the solution due to non-orthogonality of the S matrix.

b) Assume that the scattered radiation follows a power law distribution, E^n, and determine the exponent, n, by a least-square fit to the remaining points (those not identified as discrete lines are uncircled in the figure).

c) We now know T_j and S_{ij} for the continuous spectrum, and thus can construct the pulse height spectrum using equation (12). By doing this, curve (b) in Figure 5.15 is obtained.

d) Subtract the scattered pulse height curve (b) from the measured pulse height curve (a) (Figure 5.15). The pulse height spectrum of the discrete lines is obtained (curve (c)). It has been observed that smaller errors are produced in calculating the discrete spectrum if the pulse height spectrum is used for subtraction of the continuum distribution rather than the photon spectrum. The oscillation produced by inverting equation (16a) to obtain the photon spectrum on subtraction would introduce a large error. The pulse height spectrum, on the other hand, has these oscillations damped out.

3. Perform a least-square analysis of the discrete spectrum and obtain the photon spectrum. This, then, permits the identification of lines and subsequent qualitative and semi-quantitative analysis.

This last step has not been included in this section. The procedure (except for half-life correction) is similar to that described above. The results of elemental identification are indicated in Figures 5.15 and 5.16.

On-Line Data Analysis

We have presented a discussion of a general method for the analysis of digital spectra. It is the analyzed data which are of primary interest to the investigator performing terrestrial exploration or some space oriented investigation. In some instances, reduced information is required a few minutes after measurement (consider, for example, the situation of an astronaut using an analytical device for sample selection). In other circumstances, such as a geochemical mapping mission, a turn-around time of a few hours for data reduction may be required. Thus, information from one traverse on the lunar surface could be used in planning the next one. Finally, data reduction in periods ranging from days to months (important for the planning and modification of plans for future missions) is required. Rapid data analysis methods are therefore a key criterion.

A system for analyzing digital data which is transmitted from a remote field site to a central computation facility has been developed. A brief outline of the conceptual design of an on-line data system, and the way information flows through the various components, follows. For additional details the reader is referred to a paper by Trombka and Schmadebeck (1969).

On-Line Digital Spectral Analysis System

I. Selection of method for sampling of the raw data.
II. Raw data acquisition and display.
  A. Real time data handling.
    1. directly into computer and display.
    2. input through punch cards or magnetic tape.
  B. Display.
    1. semi-logarithmic.
    2. linear.
  C. Data output.
    1. printed.
    2. punch card.
    3. data plots.
III. Library preparation.
  A. Measurement of standard spectra.
    1. Compensation for electronics drift, and to obtain internal consistency.
      a) gain compensation.
      b) zero drift compensation.
    2. Setting of pulse height or energy region for analysis.
      a) shift to common energy scales.
      b) resolution change.

B. Calculation of library functions.
   1. Interpolations from standard tables.
   2. Gaussian shape functions.
   3. Change of resolution.
   4. Inclusion of calculations from other special functions.

IV. Raw data preparation.
   A. Data Compression.
   B. Background compensation.
   C. Component subtraction.
   D. Compensation for detector and electronic drift.
      1. gain.
      2. zero drift.
   E. Adjustment of scales to match library standards and raw data.
   F. Data smoothing.
   G. Calculation of statistical weights.

V. Data analysis.
   A. Raw data analysis
      1. least-square.
      2. area integration.
      3. pattern recognition.
      4. differentiation.
      5. preparation of output for V—C, below.
   B. Error Analysis.
      1. statistical variance.
      2. chi-square.
      3. correlation and interference.
      4. prepare output for V—C, below.
   C. Data interpretation.
   D. Return of reduced data from computer to investigator.

The program outlined above has been implemented for use on the CDC 3200 and the IBM 360 computers. Parts V—C and V—D of the above outline depend strongly on the specific experiment to be performed, and thus must be developed as each instrumental system and mission profile is defined.

We shall now briefly summarize the data analysis flow for the system outlined above taking, as an example, a lunar mission.

In step I, the proper gain and number of digital channels needed in order to obtain a unique solution or maximum information from the measurement are calculated. These parameters are determined primarily from a consideration of the detector resolution. Next, the digitized pulse height spectrum is obtained with a pulse height analyzer and is transmitted to the computer via a number of different types of transmission links such as satellite (e.g., *S* band transmission from the lunar surface), telephone and/or microwave transmission on earth. Once the raw data are obtained, they can be displayed on a cathode ray tube and compared with other raw data or standard spectra. Other methods of outputting the raw data are available as indicated.

The numerical data analysis methods developed in this program require comparison and utilization of standard or library spectra. In step III, various methods for obtaining such spectra are available. Further preliminary adjustments of these spectra to correspond more closely to the energy scale and resolution of the detector can be made. Once the standards are prepared, we look again at the raw data (step IV). A number of options are now available to the analyst. The data can be compressed by integration or possibly by obtaining some function of the data (for example, the logarithm of the data is of interest in a neutron die-away experiment). Following this, compensation for background can be made. Also, a method for stripping a single component or groups of components is available. Correction of gain and zero drift are then made. If necessary, data smoothing can be performed. Finally, calculation of statistical variance for each channel is carried out.

The data are now ready for analysis. The various analytical procedures are outlined in step V. The least-square method has been detailed in the earlier part of this chapter. This has been the chosen method for most applications. In certain instances the integral intensity of specific, well defined lines can be used to determine relative abundances; thus the area integration option is available. In some cases it may not be necessary to break down a given spectrum into its various components to obtain information needed by the investigation. The total spectral pattern may be sufficient to indicate similarity of species, or to permit a general categorization of a given sample: therefore a pattern recognition option is available. Once analyses have been obtained, statistical error calculations and confidence limits are determined to ascertain the validity of the results.

In certain cases, steps IV and V in the outline are performed simultaneously by proper normalization of the library functions (e.g., analysis of discrete gamma ray spectra as previously described). In other cases, all that is accomplished is the elimination of the detector effects and possibly other environmental interferences. Elemental abundances are inferred from a further analysis of the true energy spectrum (e.g., the continuous gamma ray spectrum analysis described above). It then becomes necessary either to compare results with measurements obtained in the laboratory, or with theoretical calculations. The selected approach has to be detailed for each method specifically.

Summary

As the space program continues to evolve we find that proposed measurements and experiments are increasing in complexity, particularly those experiments designed to determine the composition of the bodies of our solar system. We now face problems of dealing with complex data which must be reduced to compositional information, under special circumstances, quickly. As we have indicated, the problems are large. In this chapter, we have given approaches derived from developmental work performed in the authors' laboratory at Goddard Space Flight Center. Included are analytical methods for dealing with the complex spectra that must be measured to do these compositional studies, and a systems approach for the most efficient use of these analytical techniques. The methods proposed have worked well under laboratory conditions; it is the authors' point of view that they should serve equally well in flight programs.

References

Beale, E. M. L.: On quadratic programming calculations. Naval Logistics Quart. 6, 1959.

Bell, P. R.: The scintillation process, Beta and Gamma Ray Spectroscopy. Ed. Kai Siegbaum, Amsterdam: North Holland Publishing Co., 1955.

Bennett, C. A., Franklein, H. L.: Statistical Analysis in Chemistry and the Chemical Industry, New York: John Wiley & Sons, Inc., 1954.

Heath, R. L., Helmer, R. G., Schnittroth, L. A., Cazier, G. A.: The calculation of gamma ray shapes for sodium iodide scintillation spectrometers. IDO-17017, Atomic Energy Commission, TID-4500, 1965.

Linden, D. A.: Discussion of sampling theorem. Proc. I.R.E., 47, (1959).

Rose, M. E.: Least-square analysis for angular correlation calculations. Phys. Rev. **91,** 610 (1953).

Trombka, J. I.: On the analysis of gamma ray pulse height spectra, Dissertation, University of Michigan, 1962

— Least-square analysis of gamma ray pulse height spectra, Jet Propulsion Laboratory Technical Report **32/373,** Pasadena, Calif. (1962).

— Least-square analysis of gamma ray pulse height spectra, National Academy of Sciences, NS-3107, 183—201, 1963.

— Schmadebeck, R. L.: On-line data analysis of digital pulse height spectra, Modern Trends in Activation Analysis, ed. J. R. DeVoe, National Bureau of Standards Special Publication 312, 2, 1969.

— On the interpretation of cosmic X-ray and gamma ray spectra, NASA Technical Memorandum X-63644, 1969.

— Adler, I.: Analytical methods for non-dispersive analyses in electron microprobe analysis, Adv. in Electronics and Electron Physics, New York: Academic Press 1970.

Chapter 6: Orbital and Surface Exploration Systems After Early Apollo

Introduction

In this final chapter we shall examine manned and unmanned automated missions for compositional studies of the moon and planets to be performed either on the surface or remotely (from orbit). Many of the missions to be described are in the early planning stages, and some of these are yet only conceptual in character. In some instances they will be presented here only to show the developing emphasis and thought that is being given to such programs of exploration. We shall also present views about what we believe is a reasonable methodology for a continuing viable program.

General Considerations for Geochemical Exploration Missions

We shall examine some of the principles involved in the design of a space flight experiment. We shall consider mainly geochemical type studies and look at the following parameters:

1. The Mission profile.
2. Spacecraft capabilities and constraints.
3. Sensors.
4. Data acquisition, transmission and reduction.

There is a great deal of interdependence among these factors; ideally, one should define a) the detection system to be used, b) the constraints and design of the spacecraft, and c) the mission profile that would optimally achieve the desired scientific goal. In practice one must effect compromises and take a pragmatic approach. There have been many instances in the space effort where programs have fallen of their own weight because of attempts to design experiments well beyond the state of the art. On the other hand, there have been temptations to wait for the next step in the state of the art to develop. In some instances there has been a tendency to discard the normal, good practices of the laboratory in the performance of experiments in space flight application (the glib assumption being made that one can do better under the difficult conditions of remote analysis than in an earth-based laboratory under far more optimum circumstances).

It is the authors' point of view that one must be prepared to stop and utilize the best existing capabilities in order to be prepared for space flight opportunities as they arise. One important requirement is to make a realistic evaluation of the capabilities and potential of the proposed experiment.

Ordinarily, the mission is not designed particularly for the accomplishment of one specific objective; the economic investment is usually too great. Further, because of complexity in the design and control of space flight missions, designs must be frozen and instrument packages must be delivered anywhere from months to years before the flight. Last-minute changes and redesigns are extremely difficult to achieve. During these early periods in the development of space-science geochemistry exploration missions, myriads of plans for large numbers of different missions have been proposed; few have been implemented in short periods of time. This puts a burden on those responsible for the development of such scientific instrumentation to design instrumental systems in parallel with the development of a flight mission. The instrumental systems, as we have stated, must reflect the most advanced state of the art and must be designed, as well, to minimize the engineering interface requirements with the spacecraft. This latter requirement makes it possible to adapt an experimental package for use on a wide variety of spacecraft. Additionally, one can better respond to spaceflight opportunities given short lead times. Finally, with a simplified spacecraft-instrument interface, the lead time for the delivery of flight configured instrumentation, in response to mission requirements, can be greatly decreased. This approach has been used successfully in the development of programs of geophysical and astronomical investigations, and it is the feeling of the authors that a similar approach is necessary for missions involving compositional study.

Orbital Missions

The most promising imminent orbital missions that will include remote sensing experiments are those planned for lunar and Martian exploration. The possibilities for performing lunar orbital science are receiving increasing attention. The present indication is that orbital type exploration missions may bracket the total lunar program in terms of time. Under consideration are scientific studies to be performed from different platforms such as the Command and Service Module (CSM) orbiting the moon as part of a manned Apollo mission, instrumented subsatellites carried to the moon and then injected into lunar orbit from the CSM, and finally unmanned direct launches.

Critical Parameters for Lunar and Martian Investigations

The general results expected from a remote sensing orbital mission are twofold: global maps of elemental composition of the major constituents, and, similarly, global maps of natural radioactivity.

Consider the moon. Our viewpoint concerning it was until recently based entirely on terrestrial analog. It is considered highly likely that the bulk composition of the earth is divided, as a result of differential phase separation, into three main zones, the core, the mantle, and the crust. The high temperatures leading to this gross differentiation have been produced by the radioactive decay energies from ^{40}K, U, and Th, and the radioactive daughter products of U and Th. These elements not only are responsible for the differentiation process but also, in common with a number of other elements, are themselves differentiated. The acidic rocks typical of the earth's crust tend to concentrate the K, U, and Th so that the concentrations are as much as two orders of magnitude greater than in the original materials.

There is, however, disagreement about whether the moon has undergone substantial melting at any time in its history. Evidence can be cited to support hypotheses of extensive, moderate or minimal differentiation. The analysis of Turkevich et al. (1968) on the lunar surface and Vinogradov et al. (1967) from orbit has given us our first information. The Turkevich data, for at least three points on the lunar surface, have fixed the composition in the general range of basalts or basaltic achondrites (partially differentiated rocks). The results of the preliminary analysis of the Apollo 11 returned lunar samples indicate, at least for Mare Tranquilitatis, an elemental composition like that in terrestrial igneous rocks. There are however some surprising and significant differences in the enrichment of the refractory elements such as titanium and zirconium, and a depletion of the alkalis and volatile elements. An equally surprising result is the low potassium to uranium ratio.

While these findings give strong indications of heating and fractionation, they raise the question of a global phenomenon. Further, they point to a fractionation mechanism different from that known on the earth. Thus a total or global picture obtainable from a successful orbiter mission becomes highly significant.

We shall now examine the use of remote orbital sensing for obtaining data relevant to the solution of the questions posed above. The problems in performing compositional analysis are complex. The principal problem is to learn what observational phenomena can yield unique solutions in terms of elemental identification and concentration. A glance at the electromagnetic and particulate spectra finds a limited number of possibilities. An evaluation of these shows that none yields very precise results by laboratory standards. It is possible nevertheless to establish a heirarchy of experiments; and those at the top of the list, such as gamma ray or X-ray experiments, can provide much useful information particularly when integrated with additional measurements of backgrounds and particle emission. Let us therefore now look at the radiation environment of the moon, as it bears on the problems described above.

Gamma Rays

First, and probably most important, is the gamma ray emission of the naturally radioactive elements. Table 6.1 lists the principal gamma ray energy emission for K, U, and Th. These emitters are of major interest, and are most likely the sources of the natural activity detected from lunar orbit. Emission rates for these gamma ray lines are proportional to concentration. The most prominent line of ^{40}K at 1.46 MeV will most likely be the strongest discrete line in the spectrum, except for the 0.51 MeV annihilation line which yields little compositional information. K, U, and Th are, as stated previously, important indicators of magmatic differentiation.

Table 6.1. *Energies and relative yields of natural gammas*

Constituent	$T_{1/2}$	Range of abundance in earth's crust	Yields of gamma rays with $E>0.3$ MeV	
			Gamma energy	Yield
^{40}K	1.3×10^9 yr	$1-6\times10^{-6}$ per gram	1.46 MeV	11%
			0.35 MeV	44%
^{238}U plus decay products	4.5×10^9 yr	0–5 ppm	1.76 MeV	19%
			1.12 MeV	20%
			0.61 MeV	45%
^{232}Th plus decay products	14.1×10^9 yr	0–20 ppm	2.62 MeV	100%
			1.64 MeV	13%
			1.59 MeV	12%
			0.97 MeV	18%
			0.91 MeV	25%
			0.34 MeV	11%
			0.73 MeV	10%
			0.86 MeV	14%
			0.53 MeV	83%
			0.51 MeV	25%

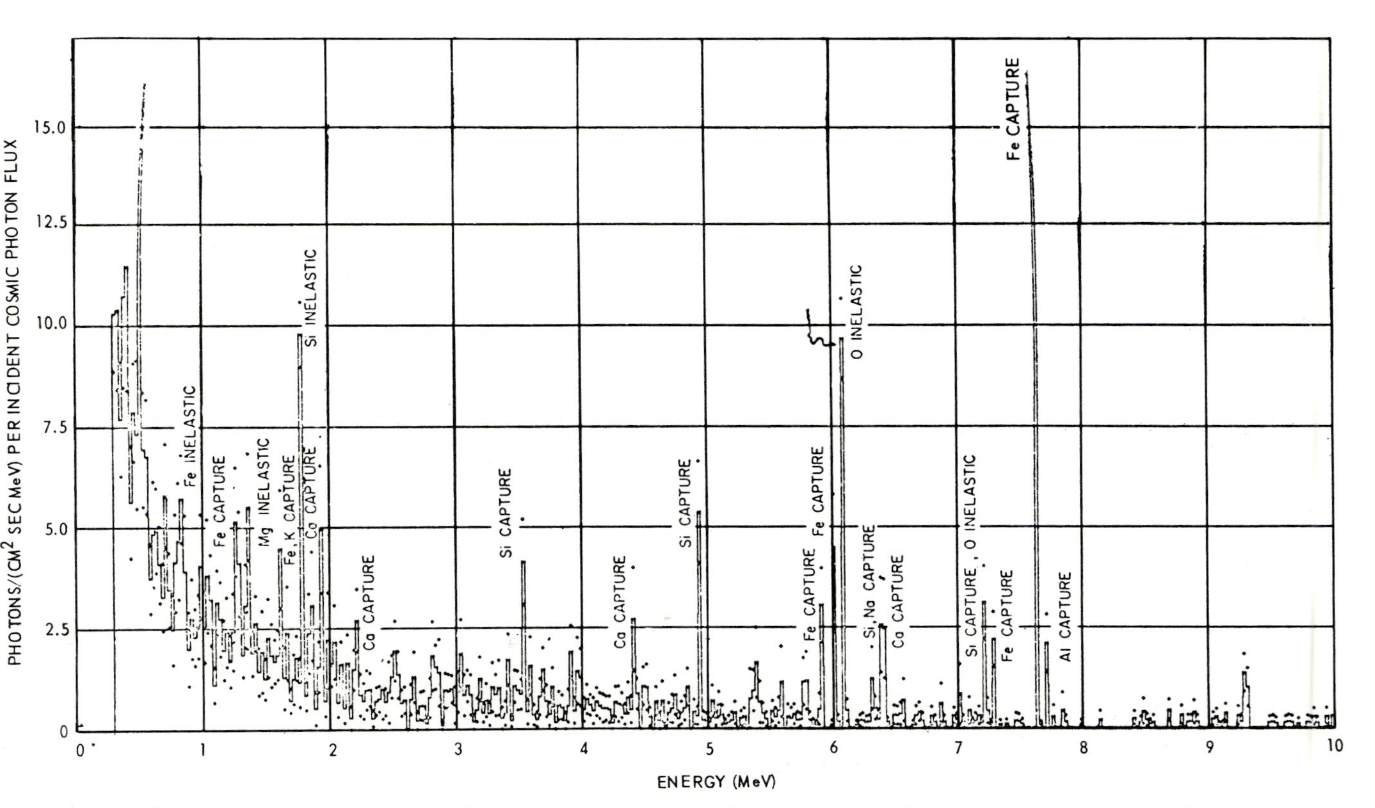

Figure 6.1 Leakage spectrum from lunar surface for basaltic composition. (After Armstrong et al., 1968)

An additional expected source of gamma radiation is that produced by cosmic ray bombardment of the lunar surface. This type of interaction produces both a discrete line spectrum as well as a continuum. The line spectrum can be used to infer compositional information. There are a number of reactions producing the discrete line spectrum. The major mechanisms are listed below:

1. Neutron capture, e.g. $p \rightarrow n$ reactions followed by neutron absorption.
2. Neutron inelastic scattering.
3. Proton and neutron activation.

Theoretical calculations of the induced gamma ray spectra have been performed by Armstrong and Alsmuller (1968). Their results, reproduced in Figure 6.1, are based on a lunar surface of basaltic composition. Using this assumption, one would expect to observe prominant lines of Fe, O, Si, Ca, and Mg. These are the most abundant elements, and their concentrations, or, more importantly, the relative concentrations (i.e., Fe/Si, Mg/Si, Ca/Mg), are of major importance in understanding the differentiation process.

Alpha Particle Emission

The next important process to be discussed involves the emission of alpha particles from the lunar surface, and the associated gamma ray activity. There are four possible sources of alpha radiation from the moon (Carpenter et al., 1968):

1. The alpha-radioactive decay of the uranium and thorium impurities in the upper few milligrams of the lunar surface materials.
2. The alpha-radioactive decay of radon and thoron (and their daughter products which have diffused out of the first few meters of lunar soil.
3. The alpha radioactivity induced by the interaction of galactic cosmic rays with the lunar surface materials.
4. The evaporation of protons and alpha rays caused at times of solar flare by the interaction of flare-associated energetic solar protons and alpha particles with nuclei of the lunar surface constituents.

Of these four sources, the second and fourth are expected to be most intense. Let us consider the effect of the phenomenon described above in source 2 on the interpretation of the gamma ray spectra obtained from the natural radioactivity.

A source of the discrete lines appearing in the natural radioactivity gamma ray spectra should be seen in those areas where subsurface gas, including radon (3.8 day half life) and theorem (55 sec half life), exists. Radon, because of its longer half-life, has time to diffuse to great distances over the surface before decaying, unless it is trapped by adsorption near its source (Kraner et al., 1966). Thoron, because of its very short half-live, will certainly decay near its source. In both cases, the daughter products not lost by recoil to space will be fixed very close to the surface. The measurement of these gases and their sources is an important objective in itself. Of possibly greater importance is the location of these sources, as they are likely to confuse the picture of the distribution of those natural radioactive elements being sought as signatures of magmatic differentiation. Areas in which there is a concentration of such gases will disproportionately indicate high concentrations of uranium and thorium compared to potassium.

The result of source 4 listed above is the presence of alpha particles with discrete kinetic energies emitted from the lunar surface. The detection of these particles will enable us to determine the nature of the phenomena producing them.

X-*Rays*

Absorption of solar X-rays by the lunar surface should give measurable fluxes of characteristic X-ray lines from the low atomic number elements excited by the X-ray fluorescence process. Under normal sun conditions, one would hope to see the K alpha lines from the more abundant elements such as oxygen, sodium, aluminium, magnesium, and silicon and the very soft L alpha lines from iron, calcium, and potassium. The more easily observable K spectral lines of iron, calcium, and potassium would be produced during solar flare outbursts. Figure 6.2 shows a comparison of the solar X-ray spectrum between 6.3 and 20 Å obtained for a quiet sun and on the following day during flare conditions. Figure 6.3 shows an observed X-ray spectrum in the region of 1.3 to 3.1 Å during in increasing phase of an X-ray outburst.

Although it is reasonable to expect qualitative information from the analyses of X-ray spectra excited at the lunar surface, quantitative data will be considerably more difficult to obtain. The characteristic fluorescence X-ray spectrum produced must be superimposed on a background resulting from scattered solar flux. This solar flux is known to contain a number of lines characteristic of highly ionized silicon and iron (Neupert, 1968). Under active sun conditions, the lines become much more pronounced (see Figures 6.2 and 6.3).

The yield of characteristic X-rays from the surface of the moon depends on its chemical composition and the nature of the exciting solar X-ray spectrum. Gorenstein (Carpenter et al., 1968) has calculated the expected yields of characteristic fluorescence X-ray lines for a typical basalt and granite. The solar X-ray flux used in the computations was taken from the OSO-3 daily averages for April 1967. A continuum non-

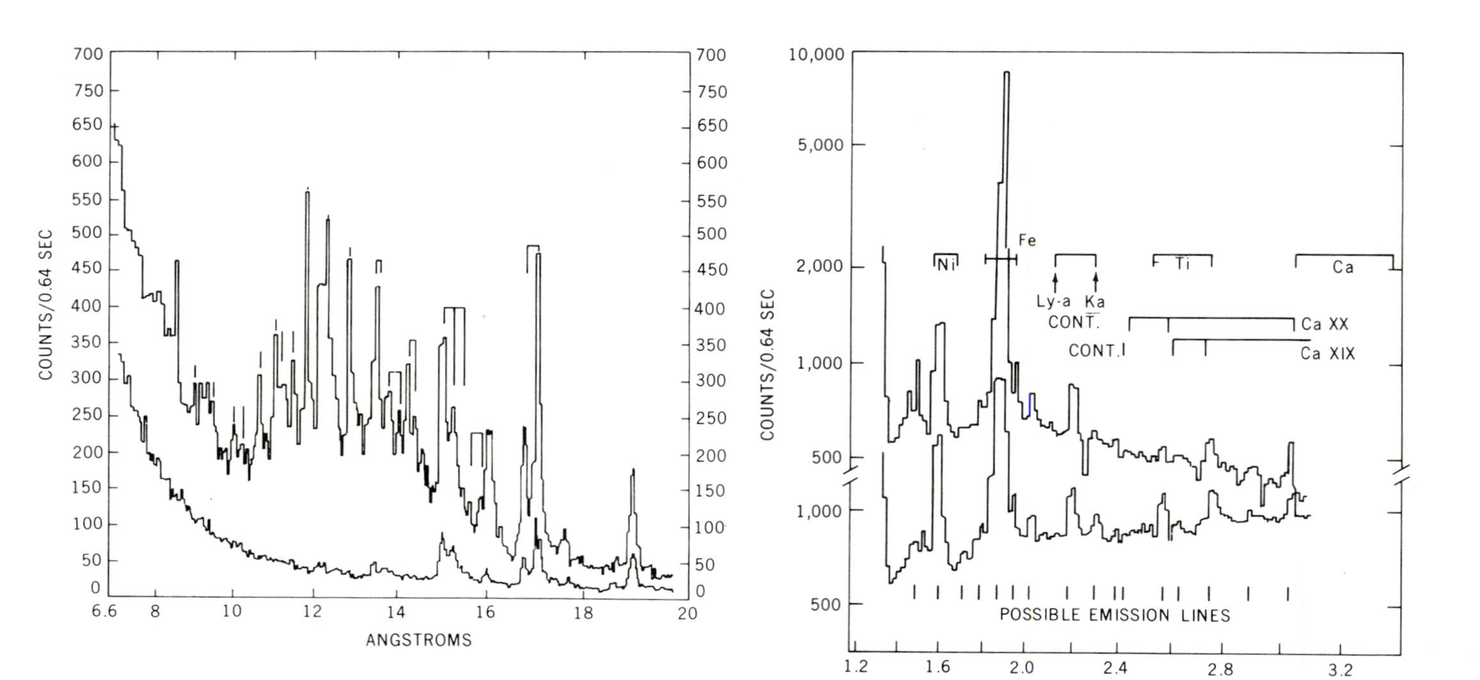

Figure 6.2 Comparison of solar X-ray spectrum between 6.3 and 20 A obtained during a flare on March 22, 1967 with a spectrum obtained on the previous day when no flares were in progress. Note: Spectral resolution is insufficient to allow resolution of lines within each array. (From article by Neupert, Gates, Swartz and Young, 1967)

Figure 6.3 The solar X-ray spectrum in the region 1.3–3.1 A as observed by Neupert, Gates, Swartz and Young during the increasing phase of an X-ray burst on March 22, 1967. Note: Apparent differences in spectral distribution are due to the increase in intensity of the X-ray burst in the time (5 min) required to make the two scans

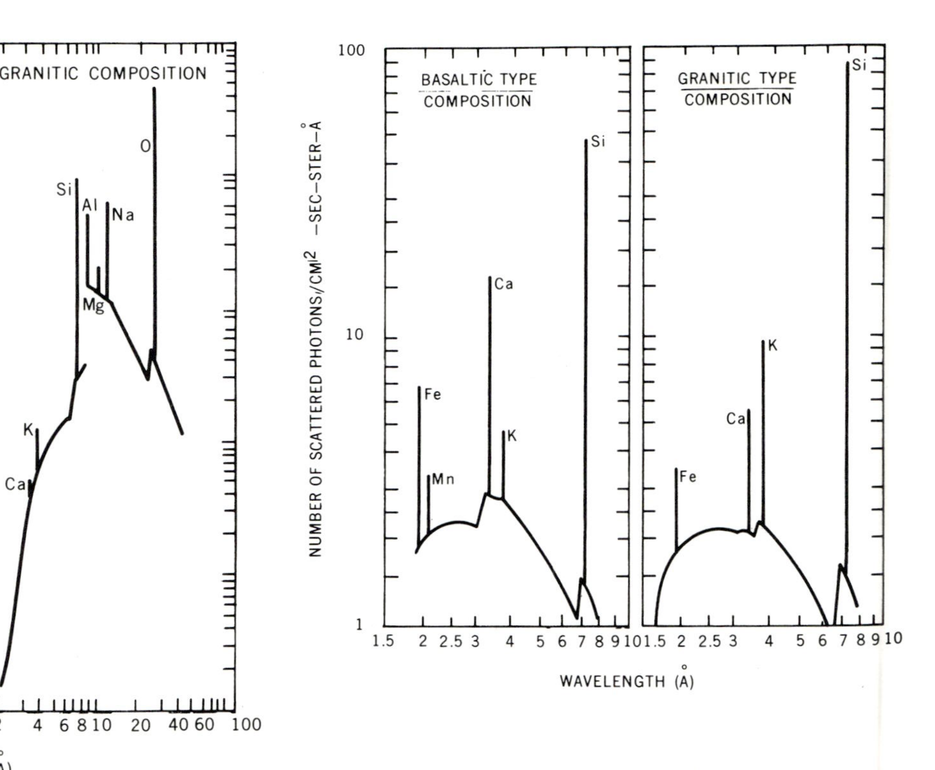

Figure 6.4 Calculated yield of fluorescent $K\alpha$ and scattered X-rays from ideal smooth lunar surface material of basaltic and granite composite. Note: The incident solar spectrum is presumed to be a free-free emission continuum characterized by $T = 6 \times 10^6$ °K whose total integrated intensity between 2 and 8 A and 8–20 A is equal to the mean of the OSO-3 daily averages for April 19. The curves represent the back-scattered solar flux

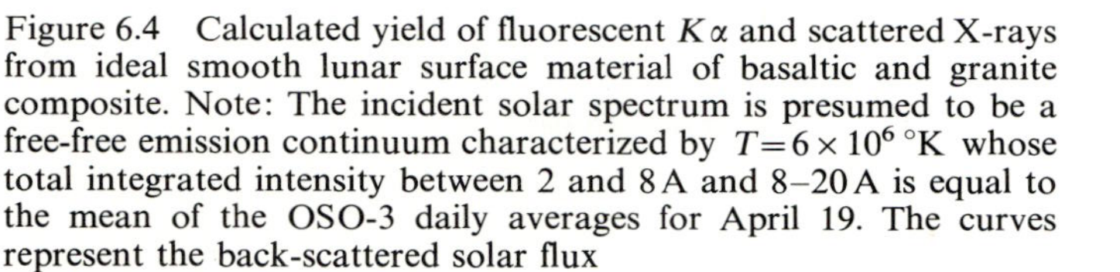

Figure 6.5 Calculated fluorescent and scattered X-ray intensity during typical solar flare (presence of lines in solar flux spectrum is neglected)

flare solar flux was assumed, and the spectral shape was characteristic of a free-free emission with $T = 6 \times 10^6$ K.

Figures 6.4 and 6.5 show the theoretically calculated spectra for a basalt and granite under two different solar conditions, and based on an idealized smooth lunar surface. The yield of characteristic K lines for potassium, calcium, and iron under the quiet sun conditions is seen to be very small (Figure 6.4). The expected yield of these lines, calculated for a typical solar flare, is shown in Figure 6.5.

Neutron Albedo

Another component making up the lunar radiation environment is the neutron flux emitted as a result of incident cosmic ray bombardment. The magnitude, energy, and spatial distribution of neutrons generated by cosmic rays have been inferred from knowledge of neutrons generated in the earth's atmosphere (Ligenfelter et al., 1961). The production rate of neutrons at the lunar surface is expected to be higher than in the

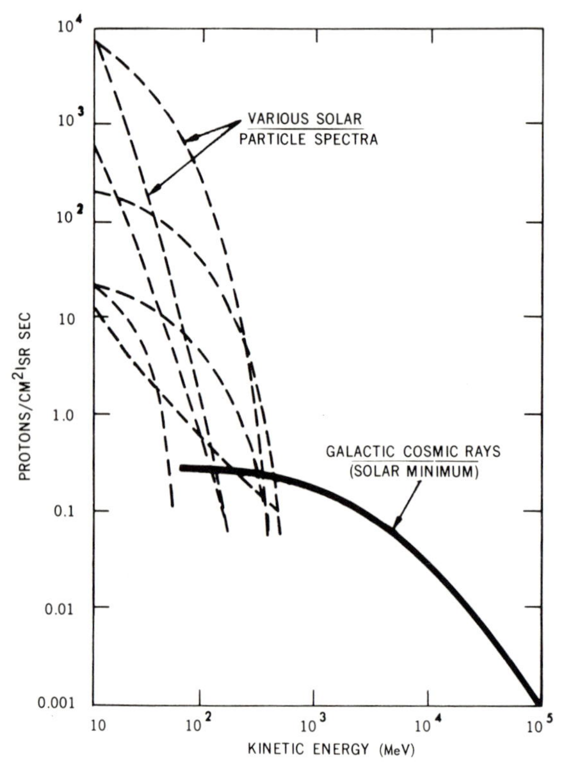

Figure 6.6 Solar and galactic cosmic ray proton flux. (Fichtel, Guss, and Ogilvie, 1963)

earth's atmosphere for the following reasons: the higher average cosmic ray intensity at the lunar surface, the higher average atomic mass of lunar material, and the added neutron production due to pi meson decay. Figure 6.6 shows the measured proton flux from the sun and galactic protons. The absence of an atmosphere and a magnetic field allow these strongly interacting particles to strike the lunar surface. This absence of atmosphere and appreciable magnetic field leads to a situation quite different from what one finds at the earth's surface. Charged particles of low momentum are not magnetically deflected, and strong interactions of the primaries take place directly with the surface materials. These interacting primaries undergo a complex series of nuclear reactions that ultimately yield several decay products. Included are such reactions as meson production, "knock-on", and evaporation mechanisms. Inelastic processes produce excited nuclei which subsequently emit gamma rays, charged particles, and neutrons. The alpha particle and gamma ray emission was discussed above.

The equilibrium neutron leakage spectrum at the lunar surface, and the neutron capture as a function of depth beneath the surface, have been calculated for a number of assumed lunar surface compositions. The results of these calculations are presented in Figures 6.7 and 6.8. The integrated leakage spectrum as determined from the above figures has been found to be very sensitive to the hydrogen content. For example, 35.9% of the neutrons leak into space for an assumed chondritic composition. This leakage decreases to 30.3% with the addition of 0.1 hydrogen atom per silicon atom, and 17.3% with 1% hydrogen atom per silicon atom. Thus, if the integrated neutron leakage fluxes from the moon can be determined, information concerning hydrogen content can be inferred. There is, however, a complicating factor that must be considered. The leakage spectrum is affected by the macroscopic cross-section for neutron absorption. It is therefore essential to determine the abundances of those elements that contribute appreciably to this cross section. One example is iron.

Infrared

Spectral analysis of absorption and reflection of infrared radiation has proved useful in helping to characterize the composition of minerals. Such measurements, however, require carefully prepared samples for precise identification and are therefore of limited usefulness for remote sensing measurements of the lunar surface.

The infrared spectrum emitted from mineral samples consists of the blackbody radiation corresponding to the sample temperature and superimposed bands characteristic of the sample material. One such band is the fundamental O—Si stretching vibration. This band has been

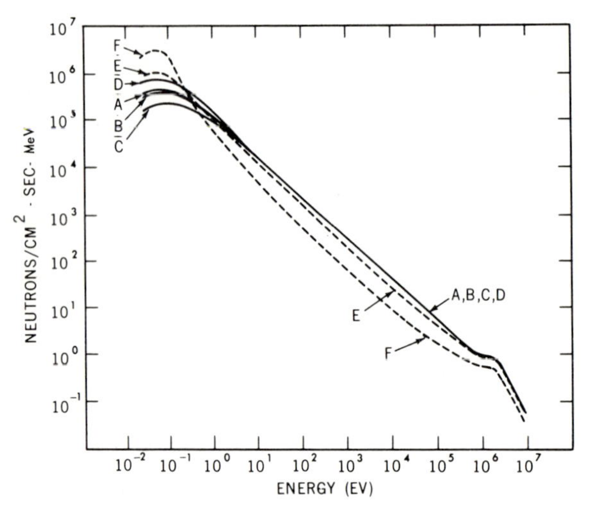

Figure 6.7 The calculated neutron equilibrium leakage spectrum for the lunar surface for compositions A, chondritic material; B, chondritic material with a 10% increase in the total $1/v$ capture cross section; C, with a 50 % increase in $1/v$ capture; D, with a 35% decrease in $1/v$ capture; E, chondritic material with 0.1 H/Si atom; and F, with 1.0 H/Si atom. (After Ligenfelter et al., 1961)

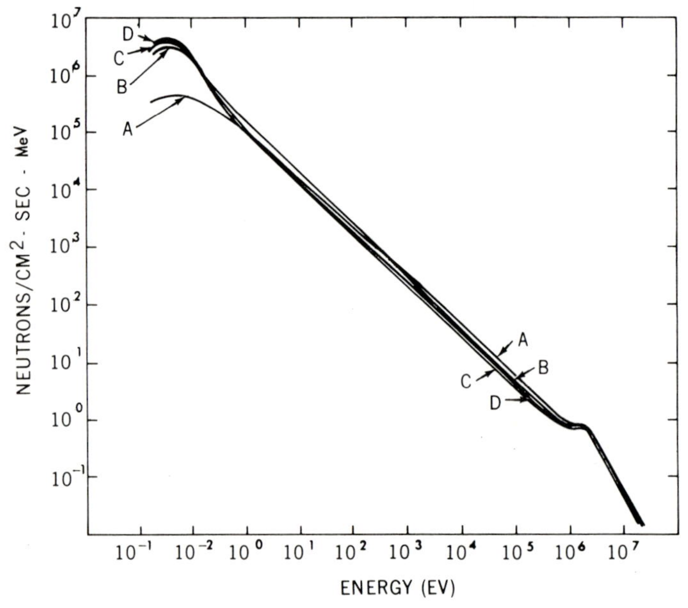

Figure 6.8 The calculated neutron equilibrium leakage spectrum for the lunar surface for compositions A, chondrites; B, tektites; C, basalt; and D, granite. (After Ligenfelter et al., 1961)

shown to shift from about 9 microns for acidic minerals like quartz, to near 12 microns for ultrabasic minerals. Thus, the position of this band, if observable, can determine the lithologic type of the rock although the spectral position of the bands are difficult to locate with accuracy. The O—Si—O bending modes appear in the 16 to 40 micron range, and their spectral positions also change with mineral composition but in a less ordered manner.

Since the moon has no atmosphere, it is an ideal object for radiometric observations; but, for the reason outlined above, it is uncertain that compositional mapping with infrared emission can be successful. It has been demonstrated both experimentally and theoretically that spectral information begins to disappear as the samples become very finely divided. As a consequence, interpretation of infrared data is hindered by uncertainties of particle size.

Background Measurements

In the above sections, radiation sources were identified which, if measured, could be used to determine the elemental composition of planetary surfaces. There also are sources of radiation which potentially can interfere with such determinations and which must either be eliminated or accounted for when making compositional studies. A partial list is given below. This list, while not complete, does reflect the radiations most commonly found in the orbital environment.

1. Charged particles: The charged particles in the cosmic rays are themselves capable of being detected by the gamma ray detector.

2. Gamma rays: Cosmic rays can serve as an exciting source for characteristic gamma ray lines from the lunar surface material. The primary cosmic rays will also produce lines from the detector and spacecraft materials that can interfere with surface measurements.

A second possible source of gamma ray interference can be radioactive sources which, for a variety of reasons, are frequently included aboard the spacecraft. This problem has plagued many a space flight experiment. A partial list of areas and components where such radiation sources have been found is listed below:

a) Radium and ^{60}Co, used to dissipate space charge effects in accelerometers and radiometers.
b) Thorium, used in lenses in optical systems for matching indices of refraction.
c) Thoriated magnesium, in structural members of the spacecraft. ^{137}Cs and ^{60}Co, used as radiation monitors for fuel gauging at zero g.
d) Promethium and other charged particle emitters, used to cause optical fluorescence in dials, door latches and target adapters.

e) Depleted uranium, used as ballast material to help adjust the center of gravity of a spacecraft.
f) Radioactive materials used as tracers for quality control by some manufacturers. These materials find their way into a spacecraft via the electronic equipment.
g) Nuclear power sources.

3. X-rays: The most significant background accompanying the excited fluorescence X-rays is the solar X-rays back-scattered from the lunar surface. The spectrum of these X-rays is derived from the incident solar flux, and yields little or no information about the elemental composition of the planetary atmosphere or surface. The presence of strong lines in the solar spectrum would complicate matters considerably if they could not be resolved from the lines being sought because of limited instrumental resolution. Solar spectrum measurements under quiet sun conditions do indeed show line emission, but the line density is high enough to allow its approximation by a continuum distribution. Flares which have been reported to exhibit strong line emission characteristic of highly ionized iron are an exception.

4. Neutrons: The mechanism for the production of neutrons in the planetary surface or atmosphere is the same as that for the production of neutrons in the spacecraft (as discussed above in the section on gamma rays).

5. Solar Wind and Trapped Radiation: Any discussion of background radiation must consider both the solar wind and trapped particles. The solar wind consists, in large part, of a flux of low energy protons and electrons (1 kev or less). These can give rise to low energy radiation from the lunar surface. Additionally, the moon traverses the tail of the earth's magnetosphere once a month and is exposed to trapped radiation (e.g., electrons and protons with energies ranging from kev to MeV.

Methods for Dealing with Background Radiation

In practice, there are various effective means for dealing with background radiation. Where charged particles are the problem, one can use active shielding such as phoswitch or anti-coincidence systems. These were described in greater detail in Chapter 2. This type of shielding is limited, however, and one also must employ analytical means for additional correction. Passive shielding (using lead or some other high density material) as a means for isolating the detectors from background radiation is impractical for the gamma ray experiment because of the production of secondary gamma radiation by the cosmic rays in the shield itself, and also because the mass of shielding is prohibitive for many flight applications.

Spacecraft cleanliness is of utmost importance, and detailed radiation surveys will be required if the remote sensing experiments considered in this section are to be carried out effectively. In some instances it will be necessary to extend an experiment away from the mass of the spacecraft by means of an extensible boom.

Finally, background measurements must be made away from the lunar or planetary surface under investigation in order to determine the magnitude and spectral composition of the background radiation. These measurements usually can be accomplished during the flight on the way to and from the planet. It is equally important, as in the case of X-ray measurements where excitation is by solar flux, to continuously monitor this incident solar flux in order to determine the appropriate background corrections.

Expected Results from an Integrated Experiment

In order to accomplish some of the foregoing geochemical objectives, it is believed that an integrated radiation package in orbit about the moon will most efficiently and uniquely determine the parameters from which major elemental composition can be inferred. Both instrument design and data analysis procedures are being considered at NASA's Goddard Space Flight Center in cooperation with experimenters from the University of California at San Diego, Jet Propulsion Laboratory, American Science and Engineering, and NASA's Manned Spacecraft Center. The instruments under consideration are a gamma ray spectrometer, an X-ray spectrometer, and an alpha particle detector.

For purposes of discussion of the mission profiles described below, we shall assume an instrumental mix as shown in Figure 6.9. Included in this grouping are gamma ray, X-ray fluorescence, and charged particle and neutron measurements. Infra-red measurements are shown as a possible option. Included in Figure 6.9 is an outline of the expected results. As can be seen, some of the measurements bear on more than one process and, in turn, other measurements either overlap or give complimentary results.

Interdependence of Measurements

Consideration given to an integrated experiment must allow for the interaction of individual measurements. If the experimental results are correct then, in general, there should be good agreement concerning the elemental abundances as well as rock types where an overlap exists. Figure 6.10

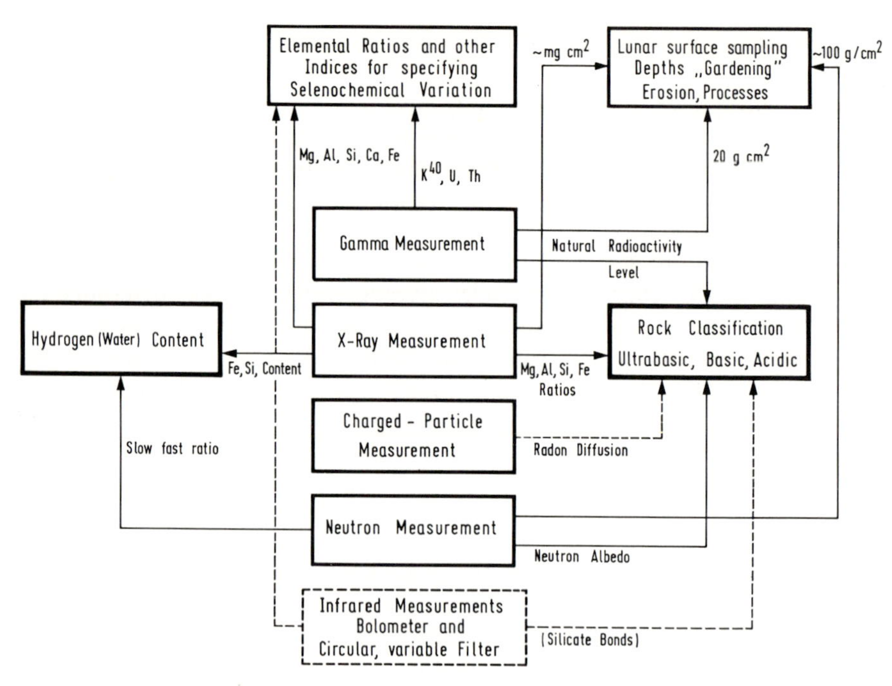

Figure 6.9 Expected results from an integrated radiation detection, remote sensing geochemical experiment

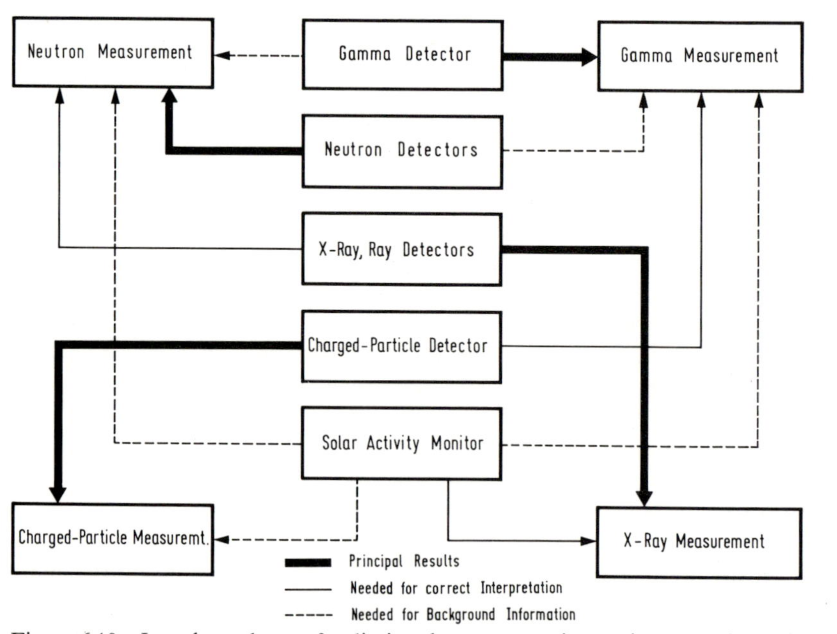

Figure 6.10 Interdependence of radiation detector experiments in terms of chemical compositional analysis

shows this interrelationship. Indicated are: the principal results, the data required from one experiment needed for the correct interpretation of the results of another, and those measurements needed for background corrections.

Orbital Missions

In the preceding sections, we have discussed the instrumental techniques and scientific results to be obtained from remote, orbital sensing missions. Both manned and unmanned missions have been considered for accomplishing these ends. There are various orbital missions possible, but we shall consider only a particular type of mission in this section.

Apollo Command-Service Module

The Astronauts on their trip to and from the moon occupy the Command Module. The Service Module remains connected to the Command Module until just prior to reentry into the earth's atmosphere (at which time the Command Module separates). Thus the Command Module is the only portion of the spacecraft that returns to earth.

Conceptually, there are two possible major modes for performing a remote sensing orbital geochemical mission. These options depend upon the method of carrying the experimental package to the moon. The instruments can either be mounted in the Service Module itself, or on a satellite which could be carried aboard the Service Module and then injected into orbit at some time after the initiation of the flight or directly launched as an unmanned mission. The pieshaped sector in which the sub-satellite can be stowed is an empty bay known as the first sector of the Service Module. This module, as presently designed, has this sector empty and thus provides a most reasonable location for carrying experimental packages (or a sub-satellite which, in turn, carries the experiments). The sector is approximately 15 ft. high and 5 ft. deep.

Let us now consider the relative merits of including the remote sensing elemental composition package in the Service Module as compared to a subsatellite. The service module is linked to the command module during the total time that the analysis instrumentation can be operated. The time and orbital constraints dictated by the mission limit the time of performance of the experimental measurements. Measurements can be made only while the Command-Service Module (CSM) is in orbit around the moon, and during transit from and to the earth. Because these missions are manned, one can consider only experiment times ranging from one day to approximately two weeks in the near future. (Limitations are set by such factors as functioning time of the life support systems). The CSM must be used in support of the lunar surface missions taking place simultaneously; thus, during the early phases of the Apollo program, missions will unquestionably be confined to equatorial orbits. These circumstances permit rather limited coverage of the lunar surface.

Figure 6.11 shows an arrangement of instruments mounted aboard the first sector of the Service Module, planned for Apollo 16 through 17. This represents a beginning in the performance of a remote sensing, geochemical mapping of the moon. Due to spacecraft constraints and limited times for implementation, the arrangement shown is not optimal but should make possible the acquisition of base line data to be used in planning subsequent missions.

A description of the instrumentation mounted in the service module is as follows:

1. Gamma ray spectrometer: The detector is a $3'' \times 3''$ NaI(Tl) scintillation counter with an active plastic anti-coincidence proton shield. A 512 channel analog to digital converter is employed and real

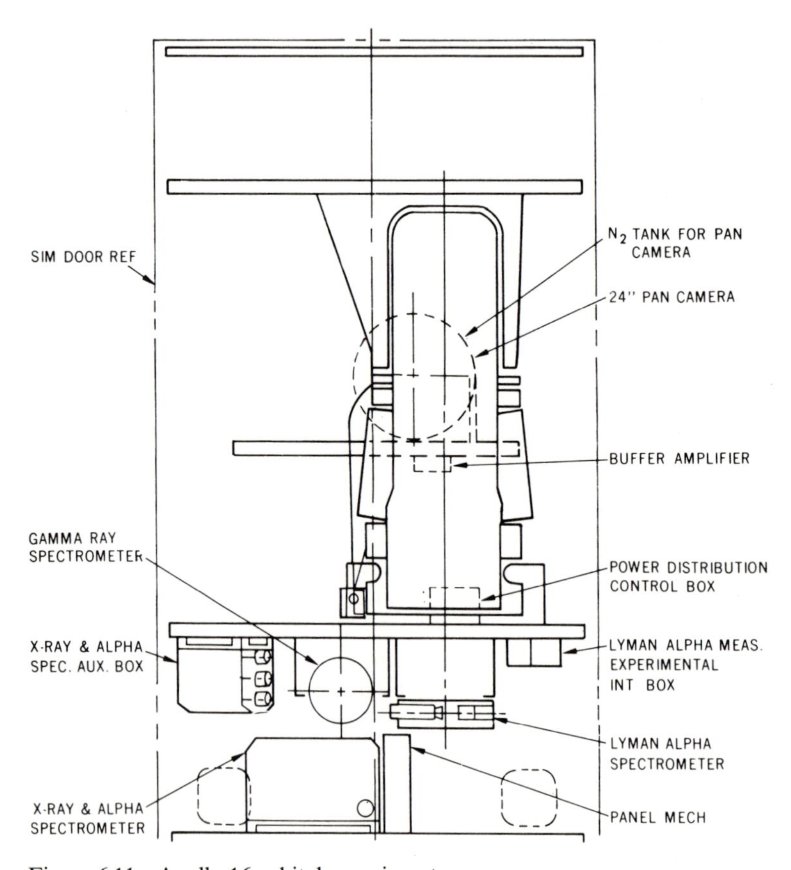

Figure 6.11 Apollo 16 orbital experiments

time transmission of the channel is performed since there is no data accumulator (memory) available for this experiment. The channel numbers, as a function of time, are recorded on tape during that period of time when the spacecraft is not in communication with the earth. The detector is extended away from the spacecraft on a 25 ft. boom in order to decrease the effect of activity induced in the spacecraft.

2. X-ray spectrometer: The X-ray experiment consists of four proportional counters mounted in the service module. Three of the detectors scan the lunar surface, and one detector functions as a solar monitor measuring the incident solar X-ray flux. The lunar detectors have an effective window area of 32 cm^2, while the solar monitor has a window area of 12.5 mm^2. Data handling involves an 8 channel stacked discriminator. Four spectra are accumulated simultaneously in a buffer memory for fixed periods of time. The accumulation time corresponds to counts taken at 8 second intervals over the total orbit. Pulse shape discrimination is used to minimize background radiation.

3. Alpha albedo experiment: The alpha detector consists of a 30 cm^2 active area made up of a composite of small, individually biased solid state silicon surface barrier detectors. The signals from each detector are coupled into respective charge sensitive preamplifiers. The output and gain of each pre amp detector combination is adjusted so that, for a given particle energy, the output from any given detector will be made nearly identical. The signals are then summed, amplified, and fed into a 256 channel analog to digital converter. Since no data accumulator is available, the method of recording and transmitting is similar to that for the gamma ray experiment described above.

The existing plans call for the above experiments to be performed after the astronauts have returned to the spacecraft from the lunar surface. The Command Service Module, and possibly the LEM, will remain in orbit for one to two days to permit the measurements. In addition to the gamma ray, X-ray, and alpha particle measurements, other experiments involving a panoramic camera and a mass spectrometer will be performed. The panoramic camera will be used for detailed photography; the mass spectrometer will study the tenuous atmosphere at orbital altitudes.

The program just described involving experiments using the Apollo Service Module should provide meaningful measurements.

Surface Missions

The major scientific objectives for surface exploration of the moon are to determine surface features and regional relationships, composition of surface and subsurface materials, the internal structure and energy budget, the interaction of the moon with the surrounding space environment, and the possible existence of organic matter. Similar objectives apply to planets such as Mars and Venus with considerably greater emphasis being placed, particularly in the Martian case, on the search for life forms or proto-genetic materials.

Surface exploration, in one main aspect, is different from orbital in that much more effort is put into the design of active experiments as opposed to the passive experiments described above under orbital missions. This means that many experiments include methods for exciting some kind of response that is then measured and interpreted. Examples of this can be cited: X-ray emission spectroscopy, X-ray diffraction, gas chromatography, various neutron gamma methods, and back scattered alpha particle spectroscopy.

A report by Hinners, James and Schmidt (1968) discusses a Lunar exploration program developed to cover the period from the first lunar landing to the mid 1970's. Their survey is based on such factors as hardware availability, increasing scientific endeavor, budgetary constraints, operational learning, lead times, and interaction with other space programs. The report divides the lunar program into four phases: 1. an Apollo phase employing existing Apollo capabilities; 2. a lunar exploration phase involving an Extended Lunar Module (ELM) with improved payloads and staytime capabilities; 3. a lunar surface rendezvous and exploration phase planned around a Lunar Payload Module landing unmanned and in support of the ELM, and permitting up to a two week mission; and, finally 4. a lunar orbital survey and exploration phase (see above).

The Apollo phase of surface exploration has begun with the first two lunar landings. The objectives of these initial missions were in a large sense exploratory from an engineering and educational point of view. Of necessity, the early mission's scientific objectives were constrained by the requirements of astronaut safety and the performance of functions necessary for the planning of future missions. Considerable time was spent learning how the Lunar Module performed and about the astronaut's ability to function in a hostile environment. From a scientific point of view, the most significant acts were the collection of lunar samples and the emplacement of the geophysical (ALSEP) experiments. Succeeding missions in this first Apollo phase will become progressively more comprehensive. Beginning with the fourth mission, one can expect mission plans to include more extensive excursions on the lunar surface. The fourth lunar landing will be made in Frau Mauro area (lunar highland). Samples from this area should provide answers to such questions as: Is there a variation in composition with respect to the mare areas and what is the local geochronology?

The following landings might well be at other, perhaps more difficult to reach, areas of high interest—a wrinkle ridge and a highland maria contact region. The major scientific activities would be similar to those of previous missions: the emplacement of a geophysical station, sample collection, and the recording of detailed observations (all activities associated with longer traverses away from the spacecraft).

As projected by Hinners et al. (1968) these early missions will yield a first cut, and therefore limited, picture of the moon's surface, composition, and structure. By the end of the early missions, there should be a useful net of ALSEP's which, during their functioning life, will supply a day by day account of the moon's internal behavior and its interaction with its external environment. We should have a reasonable picture of the structure of the upper layer of the maria, the depth of the fragmental layer, the extent of overturning and gardening, the existence of sublimates, and a very good estimate of the major element distributions. From orbital surveys we should obtain a clearer picture of the degree of homogenity of the surface, and a good estimate of the correspondence between data obtained by surface studies and those from remote orbital sensing missions.

One must now begin to examine the next step, "the comprehensive investigation phase". This final exploration phase of lunar investigation is not yet a defined scientific program. Alternatives are being considered: the termination of lunar exploration after a small number of manned landings, a pause to assess the advisability of continuing an extensive program of exploration, or a continuing program of increasing ambition. Only the last will satisfy the scientific objectives required for a thorough understanding of the moon. In the section that now follows, we will assume the third alternative as chosen, and attempt to describe planning for such missions. Parts of the program do exist, and are found in a series of recommendations made during the 1967 Summer Study on Lunar Science and Exploration held at the University of California, Santa Cruz.

In a general way, a lunar exploration phase must be built around longer stay times, more EVA's, enhanced support facilities such as Extended Lunar Modules, and increased surface mobility achieved by small flying units and/or surface rovers. Some relevant recommendations made by working groups at the Santa Cruz Conference are listed below:

1. The immediate development of a Lunar Flying Unit (LFU) to increase astronaut mobility during later Apollo flights.
2. The rapid development of a Saturn V dual launch capability.
3. The development of a dual-mode Local Scientific Survey Module (LSSM). This would be a wheeled vehicle capable of operating in an automatic or manned mode.

4. The development of an LSSM for an automated mode of operation to be used for long unmanned traverses over the moon's surface. It would be expected to perform the following functions:

a) Sample collection and ultimate rendezvous with the next manned lander.
b) Deployment of several small ALSEP-type remote geophysical monitors along the traverse route.
c) Non-destructive rock analysis en route by means of automated analytical instruments.

Of further interest are the recommendations concerning experiments and instrumentation for orbital (see section on orbital science above) and surface experiments. For surface experiments, the recommendations call for the development of the following portable devices:

1. An X-ray emission spectrometer.
2. Solids source mass spectrometer.
3. Neutron-gamma analyzer.
4. Alpha back scatter analyzer.
5. X-ray diffractometer.

To illustrate, let us consider some missions to the lunar surface as they might develop with time. The missions to be described are based on a consensus of the working groups of the Santa Cruz Conference. For most part, these again represent optimum plans calculated to give the utmost in scientific yield. The plans, as indicated, depend greatly on expanded capabilities for payload deliveries and mobilities (see 1967 Summer Study of Lunar Science and Exploration, NASA SP-157).

We begin by proposing a mission that might occur early in the post-Apollo program to central Copernicus. The logistics of this effort would involve a single launch of an Extended Lunar Module carrying two small lunar flying vehicles for increased mobility. The scientific objectives of the mission would be:

1. The investigation and sampling of a large and relatively recent impact crater site.
2. To search for samples of very old lunar rocks.
3. To obtain samples for a study of cosmic ray history.
4. To date the Copernican event and its widespread effect on the lunar surface.
5. To collect samples showing the effects of recent erosion and mixing.
6. To collect samples for shock metamorphism studies.
7. To verify the mineralogy and petrology of the central hills, including a possible dyke.

Within the above-listed objectives, the most important activities would involve sample collection, skilled detailed observations, and the deployment of an ALSEP containing (among its compliment of experiments) a mass spectrometer for in situ gas analysis.

Equipment carried by the astronaut would include containers, hand tools, and compact instrumentation. The tools would be a geological hammer, tongs, and sampling devices for incoherent material and for sections about 1 to 3 ft. in length. Among the scientific instruments would be a camera, a diagnostic sample selector (such as a portable X-ray emission spectrometer using a radionuclide excitation source), a hand carried mass spectrometer with a mass range of 2-100 for pressures between 10^{-4} and 10^{-12} torr, and an emplaced solar-cell powered mass spectrometer with a mass range of 2-50 to operate for a period of at least one year.

The next mission represents a program of greatly increased complexity involving considerably greater resources and planning, an excursion to the region near the Crater Aristarchus (Schroters Valley). A number of options are possible, but the one described here consists of a dual landing and the employment of two long range lunar flying vehicles. The total plan calls for a manned landing in the region of Aristarchus, and a rendezvous with an unmanned roving vehicle somewhere in the Mare Imbrium region.

The scientific objectives are divided into two categories, those associated with the manned exploration and those to be accomplished during the unmanned phase. Figure 6.12 an excellent view of Schroters Valley taken during a Lunar Orbiter flight, serves to set these objectives in context. The geochemical purposes of the manned landing follow:

1. Study of the Cobra Head feature as a possible site of recent volcanism and outgassing.
2. Study of Schroters Valley as a site of unusual erosion.
3. A study of the sediments in Schroters Valley for the possible existence of organic matter and ancient life forms.
4. A study of Aristarchus as a large impact crater and ejecta blanket.
5. A study of the contact between Imbrian flow material and the underlying plateau (Fra Mauro material).
6. Examination of the ejecta surface of a small young impact crater.
7. Study of the floor of a large, old, flat-floored crater as represented by Herodotus.

The primary geochemical objective of the unmanned phase of the mission is the collection of samples of various features by the unmanned rover on the way to rendezvous with the manned mission. The features to be sampled include:

1. Imbrian flows.
2. The north rim, floor, break in the west wall, and rill issuing from the Crater Krieger.

Figure 6.12 Shroters valley

3. The sequence of possible post-Imbrian sites.
4. Sections across Schroters Valley.

Also outlined are objectives to be carried out during both phases of the Aristarchus mission. A major function of the manned phase will be the collection of samples and the making of observations to supply the geological context. The existence of recent volcanic activity is one of the scientific questions, thus a mass spectrometer will be placed near sites of possible recent gaseous activity. Sample selection and packaging will be done after rendezvous with the unmanned rover. In order to perform the functions outlined for the manned landing, it will be necessary to carry the following tools and equipment:

1. Sample containers.
2. Hand sampling tools similar to the one previously described.
3. A diagnostic, portable sample analysis device such as an X-ray spetrometer or diffractometer.
4. A temperature sensor.
5. A drill corer.
6. A portable mass spectrometer for the analysis of solids.
7. A mass spectrometer to be emplaced for gas analysis.

The unmanned roving vehicle would be uniquely equipped for automated exploration en route to the rendezvous point. The contemplated payload includes: a slow scan monocular television camera, portable analytical devices (such as an X-ray emission spectrometer or diffractometer), a mass spectrometer, a sampling claw capable of taking a 250 gm sample of both coherent rock and powder, facilities for sample packaging of at least 100 sequentially taken samples, and drive tubes for cores.

A mission of the kind just described will, of necessity, require specially developed flight mobility hardware, e.g., the surface rover already mentioned. Considerable thought must also be given to communications. Monitoring must be provided to permit real time analysis of the measurements being performed by the instrumentation on the roving vehicle.

Surface Rovers

This section will describe in greater detail some of the concepts presently being developed for Lunar Rover Science. A comprehensive report titled "A Study of Lunar Traverse Missions" has been issued by J. D. Burke of the Jet Propulsion Laboratory (1968). The report covers such topics as mission criteria, instrument payloads and science criteria, baseline designs, methods of delivery, etc. Under "mission criteria", the report outlines the main functions of a long range roving vehicle. These include: 1. the transportation of instruments, the collection and identification of samples, and the telemetering of observational data to the earth while traveling fairly long distances over the lunar surface, 2. the emplacement of instrumentation along the route, and 3. the performance of experiments during pauses along the route.

The mission objectives for such roving vehicles might be geophysical or geochemical or a mix of both. For geophysical missions, one would plan on performing seismic, magnetic, and gravitational surveys along a preselected route on the lunar surface. These measurements could be done either in coordination with previously emplaced ALSEP's or simultaneously with an orbiting satellite.

A geochemical and geological exploration would call for in situ measurements, as well as sample collection, if a rendezvous with a manned mission was the ultimate goal. In the event of rendezvous, sample collection would be of prime importance. If a rendezvous were not part of the mission, then more extensive in situ measurements would be performed. Arrays of geochemistry experiments have been proposed by Burke et al. The list is extensive and includes component experiments which, at this stage, must be considered as ideas for development rather than state of the art devices. Included in the list are the following:

Natural gamma ray spectroscopy,
Neutron excited gamma ray spectroscopy,
X-ray diffraction,
X-ray emission spectroscopy,
Mass spectroscopy for solids,
Mass spectroscopy for gases,
Aluminum oxide hygrometer,
Gamma-gamma density probe,
Gas chromatography,
Alpha scattering,
Nuclear magnetic resonance.

There are many serious problems to consider in the planning of such a program. It is important, once again, to be realistic in the choice of experiments (i.e., not to base mission plans on devices that are too far beyond the state of development). It is also essential to examine the experiment critically from a scientific point of view to find if the proposed techniques, under the constraints imposed, can supply unique solutions to the problems under investigation. For some of the experiments listed above, sample preparation is crucial; successful application depends on bringing the sample to the sensor. This applies, for instance, to such devices as the gas chromatograph, the X-ray diffractometer, and the nuclear magnetic spectrometer. In other instances, the instrumental and operational complexity is such that the experiment can be considered only for advanced missions.

The concept of an automated, unmanned vehicle is an exciting one, but considerable work still needs to be done. An exciting spin-off from such a program would be the experience gained in planning science missions where the emphasis will be on unmanned probes of the planets for many years to come.

Soft and Hard Landing Probes

A possible alternative to manned exploration of the moon involves the use of unmanned probes (soft landing, such as second generation Surveyors, and hard landing probes carrying instruments to the lunar surface encased in shock absorbing material). These concepts are recommended by the working groups at the Santa Cruz Conference and have already been explored in some detail in an Advanced Studies Document 730-6 published by the Jet Propulsion Laboratory in 1967. This report considers the Surveyor the most logical choice for a soft landing vehicle, particularly because of its demonstrated success and capabilities. The spacecraft would serve as a basic bus carrying its own compliment of scientific instruments, or be used to deliver a Lunar Rover to the lunar surface. The report also includes a number of alternate science packages to be delivered to various sites on the lunar surface. Some of these (cameras, the Alpha Scatter Experiment, a touch-down dynamics and soil mechanics experiment, a surface sampler) have already been flown on the block 1 Surveyor. Other proposed experiments that have not yet been flown are: an X-ray diffractometer, X-ray spectrometer, petrographic microscope, gas chromatograph, sub-surface probe, and a lunar atmosphere experiment.

There are many advantages in using a Surveyor-like vehicle to explore the moon and planets. Such a program is likely to be less expensive and considerably less hazardous than manned flights, particularly for missions to difficult areas of the moon. Certainly the experience gained from lunar missions can be extended to flights to the near planets.

Planetary Exploration

The exploration programs just described have been mainly concerned with the moon. We shall now describe some specific efforts being directed at planetary exploration. NASA has begun the active solicitation of proposals for scientific participation in the development of future lander missions to Mars (the Viking Program). By encouraging widespread participation in the early planning stages, it is hoped that the best possible scientific devices and experiments will be developed. These developments are directed towards orbital and surface missions.

The detailed objectives for the Mars landers all bear on the questions of the origin and evolution of the planet, and the possible existence of life. Some of the prime objectives have been outlined:

1. During Entry:
   a) Atmospheric composition,
   b) Atmospheric pressure and temperature profile,
2. Landed Missions:
   a) The free and bound water in the surface,
   b) The amount and type of organic matter,
   c) The presence of biological activity,

d) High resolution photographs of the surface,
e) The inorganic composition of the surface,
f) Trace atmospheric composition and diurnal variation of atmospheric components,
g) Local meteorological information and the amount and type of solar flux incident on the surface,
h) Internal activity of the planet,
i) Surface and subsurface temperatures.

Table 6.2 is the payload suggested by the Space Board for a 1973 Mars lander, and displays types of measurements, their scientific purposes, and assigned priorities. This is, however, only a tentative list. The final instrumental package will be determined by very real mission

Table 6.2. *Suggested* 1973 *Mars lander playload in order of priority*

Objective	Priority	Functional range	Est. weight	Data bits
Imaging	1	1 gross scan upon landing, 1 high resolution of sampler	4 lb	10^7
Soil Sampler	1	Up to 1 g	2 lb	10^2
Pyrolysis (gas chromatograph/mass spectrometer	1	10 to 140 amu Dynamic range, 10^6	16 lb	10^5
Direct Biology	1	Growth/Metabolism	10 lb	10^3
H_2O Detector	1	Sensitivity, 10^{-5} mb	1 lb	10^3
Gas chromatograph	2	Atmosphere: H_2, He, CO, N_2, N-oxides, CO_2, HCN, O_2, NH_3, CH_4, C_2H_2	4 lb	2×10^3
Differential thermal analysis	2	10 mg at 0.01 % H_2O	1 lb	2×10^3
Penetrating subsurface H_2O probe	2	—	3 lb	10^3
Soil temperature	3	$\pm 1\,^\circ C$	0	10^3
Air temperature	3	$\pm 1\,^\circ C$	0	10^3
Air pressure	3	± 0.1 mb, range to 30 mb	1 lb	10^3
Element analysis	3	X-ray fluorescence X-ray diffraction	20 lb	10^4
Neutron probe	3	Permafrost (?)	3 lb	10^3
Air velocity	4	Range up to 200 km hr^{-1}	2 lb	10^3

* A 1-lb life detection experiment may be feasible.

constraints which, for the Martian flight, are rigorous. The proposed instrumental devices must be able to survive high temperature sterilization procedures, and perhaps a hard landing on the Martian surface. Present plans call for modest payloads of 25 to 50 lbs. Past experience indicates that these numbers may easily diminish to smaller permissible weights. An additional constraint will be the experiments' ability to survive extended periods of transit to the Martian surface. (For additional details, refer to "Briefing for Mars Lander Scientist Solicitation, NASA (October 1968).

References

Armstrong, T. W., Alsmiller, R. G., jr.: An estimate of the prompt photon spectrum arising from cosmic ray bombardment of the moon. Oak Ridge National Laboratory Technical Memorandum 2100, 1968.

Burke, J. D.: A program analysis for lunar exploration. Jet Propulsion Laboratory Report **760/5,** 1968.

Carpenter, J., Gorenstein, P., Gursky, H., Harris, B., Jordan, J., McCallum, T., Ortmann, M., Sodikson, L.: Lunar surface exploration by satellite. Final Report, NAS 5-11086, 1968.

Fichtel, C. E., Guss, D. E., Ogilvie, K. W.: Solar proton manual. NASA Technical Report R-169, 1963.

Hinners, N. W., James, D. B., Schmidt, F. N.: A lunar exploration program. NASA Technical Memorandum 68-1012-1, 1968.

Kraner, H. W., Schroeder, G. L., Davidson, G., Carpenter, J. W.: Radioactivity of the lunar surface. Science **152,** 1235 (1966).

Ligenfelter, R. E., Canfield, E. H., Hess, W. N.: The lunar neutron flux. J. Geophys. Res. **66,** 2665—2671 (1961).

Neupert, W. M., Gates, W., Swartz, M., Young, R.: Observation of the solar flare X-ray emission line spectrum of iron from 1.3 to 20 Å. Op. J. (Letters), 1968.

Space Science Board, National Academy of Sciences, National Research Council. Planetary exploration, report of a study, June 1968.

Stecher, F. W.: Isotopic gamma radiation and the metagalactic cosmic ray intensity. Nature, **220/5168,** 675—676 (1968).

Turkevich, A. L., Patterson, H. J., Franzgrote, E. J.: The chemical analysis of the lunar surface, 69th convention lecture. Am. Scientist, 312—343, Winter 1968.

Vinogradov, A. P., Surkov, I. A., Chernov, G. M., Kirnozov, F. F.: Measurements of gamma-radiation of the moon's surface by the cosmic station Luna 10. Geochemistry 8, V. I. Vernadsky Institute of Geochemistry and Analytical Chemistry, A. S. SSR (Moscow, USSR), 891—899, 1967.

Subject Index

Typesetting and printing: Zechnersche Buchdruckerei, Speyer
Bookbinding: Konrad Triltsch, Würzburg